Wakefield 3A

A Modern Introduction to Chemistry

A Modern Introduction to Chemistry

Second Edition

D. WHITE, B.Sc., M.A.
Headmaster of Rydens School,
Walton-on-Thames, Surrey

AND

W. P. SHAW, B.Sc.
Head of Science, Boundstone School,
Lancing, Sussex

Wheaton
A Division of Pergamon Press

A. Wheaton & Company Limited
A Division of Pergamon Press
Hennock Road, Exeter EX2 8RP

Pergamon Press Ltd
Headington Hill Hall, Oxford OX3 0BW

Pergamon Press Inc.
Maxwell House, Fairview Park, Elmsford, New York 10523

Pergamon of Canada Ltd
75 The East Mall, Toronto, Ontario M8Z 2L9

Pergamon Press (Australia) Pty Ltd
P.O. Box 544, Potts Point, N.S.W. 2011

Pergamon Press GmbH
6242 Kronberg/Taunus, Pferdstrasse 1, Frankfurt-am-Main,
Federal Republic of Germany

First published 1967
Second edition 1979

Printed in Great Britain by A. Wheaton & Co. Ltd, Exeter
ISBN 0 08 022620 5

Contents

Plates 1 and 2 are reproduced from
photographs by courtesy of
Shell Photographic Service, London;
and Plates 3, 4, 5, 6 and 7
are reproduced by courtesy of
The Science Museum, London.

Preface

to the Second Edition

The aim of the book has not changed since the first edition was published. Provision of an up-to-date introduction to chemistry for the eleven- to fourteen-year-olds is still the aim but various changes in the approach to the subject not least the introduction of SI units and the increased emphasis on safety in the laboratory have all made the publication of a new edition essential.

In the preparation of this new edition I am very pleased that Mr W. P. Shaw, Head of Science, Boundstone School, Lancing, has been able to join forces with me. We both feel that the initial approach to the subject of chemistry by a study of well-known materials such as rock salt, oil and ink with a view to introducing important techniques such as filtration, distillation and chromatography should be retained. Other chapters have been improved and brought up to date where necessary and a completely new chapter on metals has been added. The inclusion of topics such as electrochemistry, atomic structure, radioactivity and the Periodic Table proved very attractive and useful in previous editions and as children hear about these subjects often well before the age of eleven we felt it desirable to include them again.

Exercises at the end of each chapter and at the end of the book are once again designed to encourage the reader to think and not merely to reproduce the text. Numerous experiments are included to add interest and to develop the scientific approach.

Our thanks are due to many who have helped us in the preparation of the book. Unfortunately it is impossible to mention everyone, but we would like to thank particularly the staff of A. Wheaton and our wives and families for their invaluable help, support and understanding.

D. White and W. P. Shaw

1 What Is Chemistry?

Science is a vast subject and includes the study of the Universe, the stars, the planets and their satellites and everything thereon. Because this study covers such a wide field it is convenient to divide it into separate branches, and scientists usually devote their time to studying one of these branches or even, because there is so much to know, a section of one of these branches. Some of the branches you may know are as follows:

Science

Chemistry | Physics | Biology | Geology | Astronomy

but there are others such as meteorology, oceanography, seismology and archaeology. If you do not already know the meanings of these words, you can look them up in the dictionary.

Chemistry is the study of the materials which make up the Universe, and when we study chemistry we try to find out what substances are made of, how they react on heating, on cooling, on putting into water and other liquids, on subjecting them to pressure and many other processes. Then the chemist attempts to use his knowledge of this behaviour to advantage in supplying some new substance or in increasing the useful properties of a substance. Chemistry is so much a part of our lives that it is very easily taken for granted. Metals, glass, plastics, dyes, drugs, insecticides, paints, paper, fertilisers, soaps, detergents, explosives and perfumes are all made by chemists.

How did it all begin? Man has been on earth for about 500 000 years and the question is difficult to answer. It is, however, known that the Near Eastern civilisations, during the period 6000–4000 B.C., were quite skilled at pottery, metallurgy, brick making and glass making. The word "chemistry" may have been derived from the Greek word for Egypt, *chemia*.

Without doubt chemistry has been responsible for many important discoveries but there is plenty of scope for many more. By the year 2000 the world's population will have doubled, and in 100 years' time it will be six times what it is today if the present rate of increase continues. As the population increases, the demand on raw materials increases. In the past 60 years alone, more of the world's mineral deposits have been used than throughout the whole of man's previous existence, and at the present rate of consumption all the world's lead and copper mines will have been worked out before the end of the century. The substitution of comparatively rare metals by more common ones or plastics will ease the problem, but the scope for future chemists is obvious.

Coal is one of the few abundant raw materials of the United Kingdom. It can give a very warm and cheerful glow when burned, but did you realise that when heated without air it can provide us with several valuable substances? The following chart will give you some idea of the value of coal.

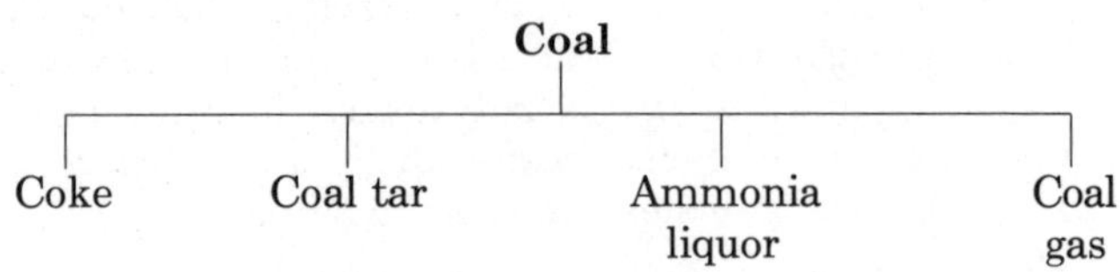

Coke and coal gas are useful fuels; coal tar is a mixture of substances which can be separated and then used to make beautiful dyes, various plastics and explosives; ammonia liquor is used to make fertilisers. Only about 12 per cent of the coal mined in Britain is used to make chemicals. The time may well come when more has to be used in this way.

Natural gas and oil found during the last decade or so under the North Sea have enriched the United Kingdom's known reserves of raw materials. Natural gas, which is mostly methane, is now piped for use as a fuel for domestic and industrial use. Oil is a valuable mixture from which petrol, kerosene, diesel oil and lubricating oils can be obtained, as well as valuable plastics like P.V.C., polythene and terylene made by chemical processes.

Many chemical changes can take place without the aid of man. The impressive formations of stalactites and stalagmites in limestone caverns; the beautiful colours of plants and animals; and even life processes such as respiration involve chemical changes. In many cases nature is more efficient than man in carrying out reactions. For instance, the octopus can extract copper from sea water to make a blue substance called haemocyanin, which is its oxygen carrier and equivalent to our red haemoglobin which contains iron instead of copper. So far, man has been unable to extract copper profitably from the sea but perhaps one day a method will be found. In the animal world chemistry is put to use as a means of protection. The "ink" ejected by the giant squid on retreat owes its dark colour to the substance melanin which, incidentally, also accounts for the colour of human skin, hair and eyes. The notorious skunk protects itself by ejecting a nasty smelling substance called butyl mercaptan to deter would-be attackers. At this point it is worth emphasising that not all chemicals have unpleasant smells; some are very pleasant indeed and include all perfumes. Haemocyanin, melanin and butyl mercaptan are all made by natural chemical reactions.

Sometimes we hear of discovery by accident. The first coal-tar dye was made in 1856 by an 18-year-old chemistry student, William Henry Perkin. In an attempt to make quinine, the naturally occurring drug used to treat malaria patients, Perkin obtained a blue dye which was called mauve. Although he did not succeed in making quinine – this was not achieved until 1945 when Woodward and co-workers in the U.S.A. succeeded – he did make the world's first coal-tar dye. As another example of chance playing its part, there is the case in 1868 when oil prospectors in Louisiana, U.S.A., sunk a well which produced little oil but did show signs of the presence of the yellow solid, sulphur, which is used to make the extremely important sulphuric acid. Surveys showed that the deposits were so great that they could make the U.S.A. independent of supplies from Sicily, the world's major sulphur producer at that time. Later you will learn how Henri Becquerel in 1896 discovered radioactivity quite accidentally; and how Marie Curie in her search for uranium discovered a new element, *radium*, which is nearly a thousand times as radioactive as uranium and more valuable than gold. The radiation emitted by radium is used today as a valuable weapon in the fight against cancer.

Chance certainly plays a part in chemistry as in all walks of life, but most of the discoveries are the result of carefully planned work and are none the less interesting for that!

THINGS TO DO

- Try to find out if there are any works in your area where use is made of chemistry. Make a list of the works and the products obtained from them.
- Make a list of things at home which have been made by chemists.
- Make a list of things made by chemists which you have noticed on your way to school.
- Try the following crossword puzzle. Copy it out in your book and if you cannot complete it now you will find the answers later in the text.

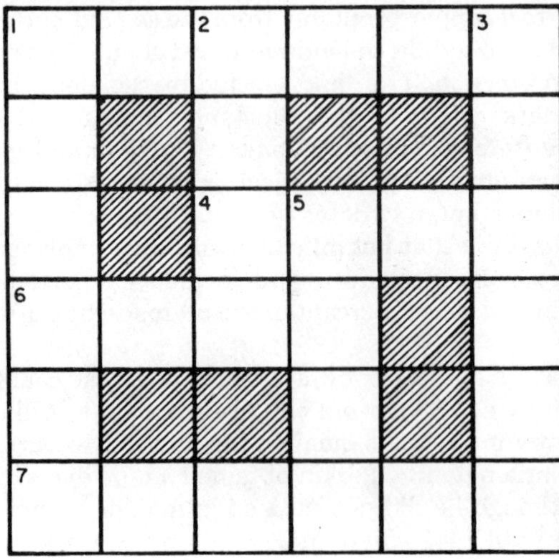

CLUES

Across

1. A metal mentioned in the Bible and once used in coins.
4. Can be dangerous; used in car batteries.
6. A–on and –st are present in the air. Which four letters are missing?
7. Gas in the air and used for balloons.

Down

1. Found in potatoes.
2. A dense metal used for plumbing and protecting roofing.
3. This kind of treatment may be used to cure some cancer.
5. Madame –e was famous for her work on radioactive substances.

2 Separating Mixtures

The separation of some substances from a mixture is very easy indeed. For example, we have no difficulty in separating knives and forks from a mixture of the two because knives and forks have characteristic shapes which are large enough for the eye to distinguish. The separation of a mixture of coke and coal is a little more difficult because there is little difference in shape and colour, but on closer inspection we find that coal is more shiny and more dense. Again we can separate the mixture by hand.

Some mixtures are much more difficult to separate. In fact, some cannot possibly be separated by hand alone. An inspection of rock salt, a mineral found in the ground, shows that it is a mixture of at least two substances – one a white substance, the other a brown substance. The rock salt can be ground up into smaller particles and separation of the white substance from the brown one would be impossible by hand alone because they are so small. Sea-water, oil, air and ink are further examples of mixtures which cannot be separated at all easily.

In this chapter we shall have a look at some of the ways we can use to separate mixtures.

Most methods depend on the following properties:

1. Solubility   2. Volatility   3. Adsorption

These three words are probably new to you but you will soon understand their meaning. We can deal with them in turn.

1. SOLUBILITY

When a lump of sugar is added to a cup of tea, the sugar disappears on stirring. What has actually happened is that the sugar lump, or even one of the grains of sugar that would make up a spoonful, is broken down into extremely small particles, so small in fact that they cannot be seen. These sugar particles are spread evenly throughout the tea. We say the sugar has *dissolved* in the tea or the sugar is *soluble* in the tea (Fig. 1).

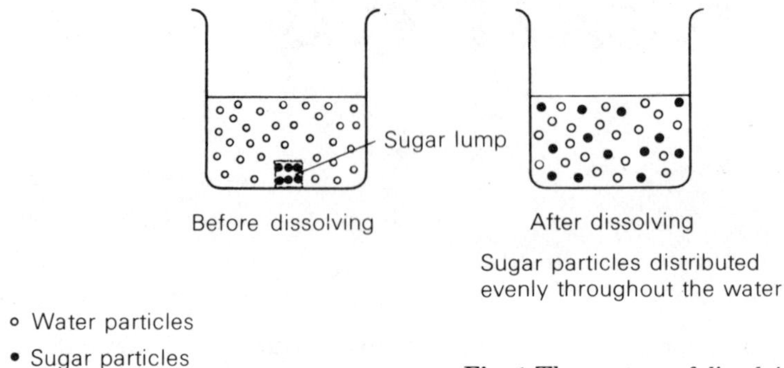

Fig. 1 The process of dissolving

Of course you will not be able to see sugar dissolving if milk is present in the tea, but by adding sugar before the milk you can see the sugar disappear or dissolve.

Other substances also dissolve in water, for in the action above it is the hot water with which the tea is made that dissolves the sugar. When a little salt is added to water in a test-tube and the tube is shaken, the crystals of salt disappear or, more strictly, dissolve and a clear liquid remains. We call the solid which dissolves the *solute* and the liquid which dissolves it the *solvent*. The clear liquid produced is the *solution*.

$$\text{SOLUTE} + \text{SOLVENT} \longrightarrow \text{SOLUTION}$$

Substances which dissolve easily are said to have a high *solubility*.

Water is an excellent solvent and most solutions you will use in the laboratory are made up with water. It is also the medium in the human body in which most life processes such as digestion are carried out, and solutions of salts in water are taken up by the roots of plants from the soil. All solutions in water are called *aqueous* solutions (Latin *aqua* = water).

Separation of the white and brown solids in rock salt

Dissolving can be a very useful tool for separating a mixture of two solids. If we consider rock salt, abundant supplies of which have been found in Poland, the U.S.A. (California), Germany (Stassfurt) and England (Cheshire), one of the striking features is that it looks very dirty compared with table salt. Separation of the salt from the dirty impurities can be achieved because the former is soluble in water whereas the latter is not. Rock salt can be heated in a beaker of water (Fig. 2), and the salt solution formed can be separated from the insoluble part by *filtration* (Fig. 3).

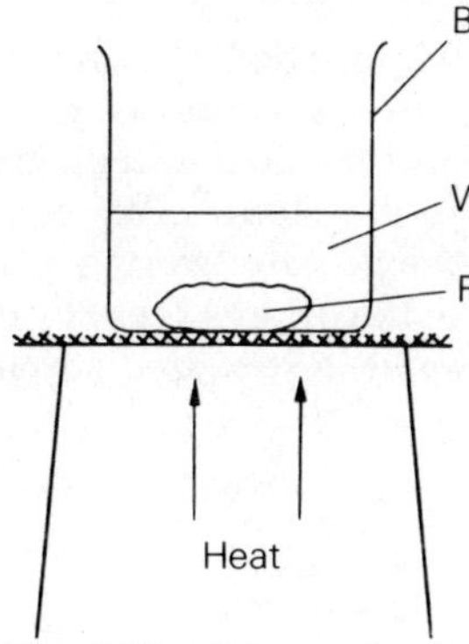

Fig. 2 Dissolving rock salt in water

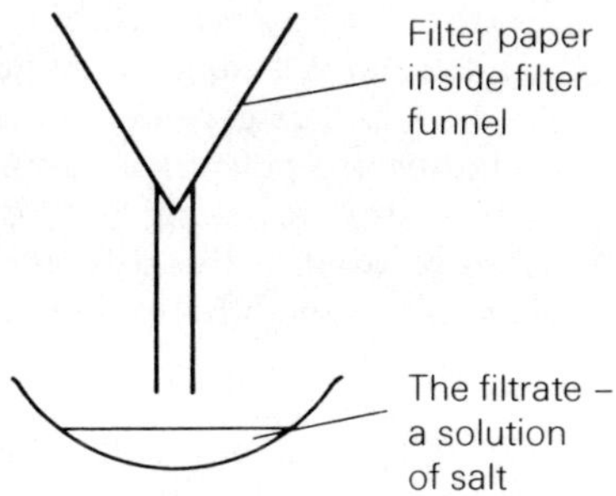

Fig. 3 Filtration

Insoluble matter will remain on the filter paper and the salt solution, called the *filtrate*, will pass through.

Can you work out a method of recovering the salt from the filtrate? If not you will find out how to do this later in the chapter.

Filtration in everyday life

This method of separating a mixture of substances is common in everyday life. A gardener uses a riddle to separate his soil from unwanted stones, while in the home his wife might be using a tea-strainer to separate the tea from unwanted tea-leaves. The milk used in the tea will have been filtered at the diary before arriving in the motor vehicle, which will have under its bonnet three filters – one for the petrol, one for the oil and one for the air needed to burn the petrol in the engine. The domestic water supply will have been subjected to filtration in the purification process. The filter tip of cigarettes removes the larger tarry particles before the smoke enters the mouth. Can you think of other uses of filtration in everyday life?

Purification of water

Water used for drinking purposes, swimming pools, etc. has to be purified and filtration plays an important part. One common method is to allow the water to seep through beds of sand, fine gravel, coarse gravel and finally bricks or large pebbles in that order. These various layers act as filters and the impurities are thus removed from the water which trickles out of the bottom.

After filtration of this sort the water is given further treatment including chlorination to kill bacteria.

The centrifuge

One of the disadvantages of separating a solid from a liquid by filtration is that the process is rather slow. The centrifuge enables us to separate a solid and a liquid very quickly indeed. But how does a centrifuge work? Let us consider the spin-drier used to dry clothes. When wet clothes are placed in a spin-drier which is then rotated very quickly, the clothes are flung to the perforated walls of the drier. As the clothes are pressed in this way against the walls, water is squeezed out and passes through holes, so drying the clothes.

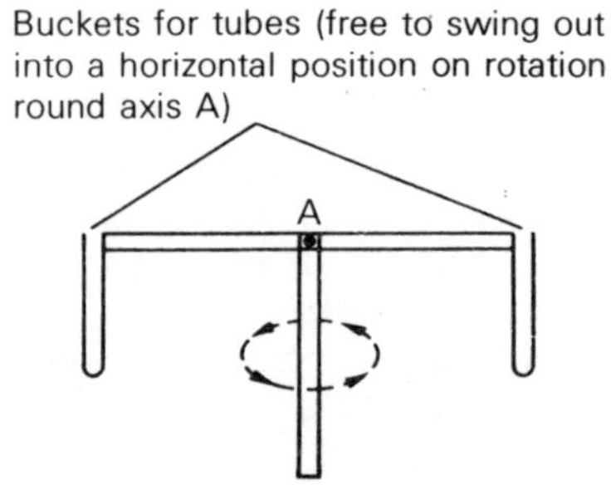

Fig. 4 The centrifuge

Using a similar principle, the laboratory centrifuge has an electric or hand-driven motor which rotates "buckets" in which glass tubes can be placed (Fig. 4). One tube contains the mixture of solid and liquid, the other the same volume of water in order to balance the two arms of the centrifuge. As the tubes are spun round they take-up a horizontal position and solids are flung to the bottom of the tube and the liquid is left above. When rotation stops, the solid and liquid remain separated and very often the liquid can be poured off leaving the solid firmly in the bottom of the tube. Try centrifuging some creamy milk and see the effect.

Non-aqueous solvents

Although water is an excellent solvent it will not dissolve everything. If you have ever helped with house decorations you will know that water will not dissolve paint. If you have done a bicycle repair or helped with a car repair you will not have removed oil or grease with water alone. Water has, in fact, many limitations as a solvent. Solvents other than water are called non-aqueous solvents. Perhaps you know what substances some of the non-aqueous solvents will dissolve. For example:

Alcohol	Used for purifying organic compounds. Tincture of iodine is a solution of iodine in alcohol.
Mercury	Dissolves silver and tin to form dental fillings.
Turpentine	"Thinning" paint.
Trichlorethylene	Dissolves grease from metal before paint spraying, e.g. in the motor-car industry.
Perchlorethylene	Dry cleaning.

Solubility curves

No one would even think of making tea by pouring cold water on to tea leaves. Apart from cold tea being far less desirable than hot tea, the very weakness of the drink would make it unpalatable. Generally speaking the hotter the solvent, the more solute will dissolve. We can illustrate this.

Fig. 5 Heating in a water bath

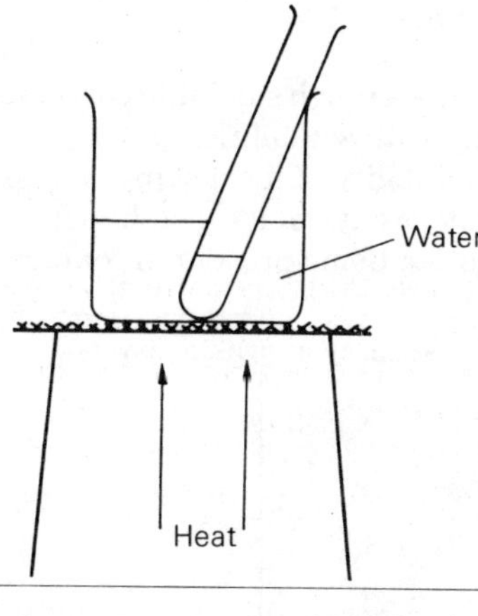

Experiment – The variation of solubility with temperature

Put 20 cm^3 water into a boiling tube and add potassium nitrate, a little at a time, with stirring until no more will dissolve. Heat the tube in a beaker of water to 30°C, the temperature being read on a thermometer placed carefully in the boiling tube (Fig. 5). As the potassium nitrate will now be completely dissolved, add more, a little at a time with stirring, until the solid is *just* in excess, i.e. until a very small amount remains undissolved. The temperature must be kept as near as possible to 30°C during the addition of the solid.

When the solid is just in excess, note the exact temperature of the solution, remove the tube from the water bath and allow it to cool.

Using five more boiling tubes each containing 20 cm³ of water, carry out the above experiment at temperatures of 40°, 50°, 60°, 70°, and 80°C instead of 30°C. When all six solutions have cooled to room temperature, the heights of the solid which has come out of solution on cooling will give a rough measure of the solubility of potassium nitrate at the temperatures used.

N.B. On cooling the hot solutions, potassium nitrate crystallises in needle-like form. It will be necessary to stir the solid so that it forms a thick solid layer at the bottom of the tube. The height can then be measured.

Temp. (°C)	30	40	50	60	70	80
Height (cm)						

Plot a graph of height of column against temperature.

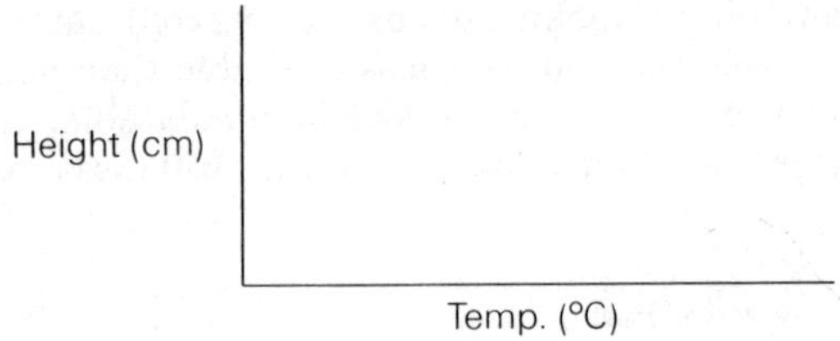

The graph is a rough solubility curve. At a later stage you will deal more accurately with solubilities.

The solubility of a substance is the number of grams of the solute which will dissolve in 100 g of the solvent at the particular temperature, excess solute being present in contact with the solution.

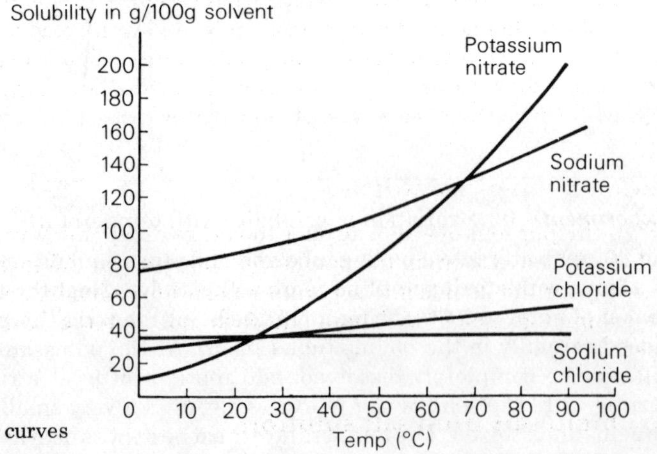

Fig. 6 Solubility curves

Examples of solubility curves (Fig. 6)

Each solute has a characteristic solubility curve. Some are steep; some are not. The shape of a solubility curve is an important factor in purifying a solid.

Crystallisation

When a solute "comes out" of a hot solution on cooling we say the solute has *crystallised.* The process of crystallisation is a very useful method for purifying a solid. In choosing a solvent suitable for crystallising a given solid, we need a solvent which will easily dissolve the solute at relatively high temperatures, but will not dissolve much at room temperature. In other words the solubility curve should be fairly steep. In the actual process, the solute is dissolved in the minimum quantity of hot solvent and, after filtration to remove insoluble impurities, the filtrate is allowed to cool. We shall study crystallisation in more detail later, but if you want to try to crystallise some solids from water the following can be used to obtain pure well-shaped crystals.

(a) alum
(b) copper sulphate
(c) lead iodide

Look carefully at the shapes of the crystals.

2. VOLATILITY

After a rainfall the pools of water do not stay on the ground for ever. They soon dry up because the water changes to water vapour which rises into the air. We say the water has evaporated.

Water is not the only liquid which will evaporate. If you have ever watched a car being filled with petrol you will have noticed the petrol fumes rising from the tank, especially on a hot day. Petrol evaporates more easily than water. We say it is more *volatile.* Liquids vary in the ease with which they evaporate, i.e. in their *volatility.*

Comparison of volatilities

Put 20 cm^3 portions of water, alcohol and ether in separate evaporating dishes and leave them in the open for an hour or two. After this short period, pour the liquids into separate measuring cylinders and record the volumes. Which of the three liquids is most volatile; which is the least?

To obtain salt from salt solution

Earlier in the chapter, when we were dealing with rock salt, we were left with the problem of obtaining salt from salt solution. We can do this in the laboratory by the process of evaporation (Fig. 7)

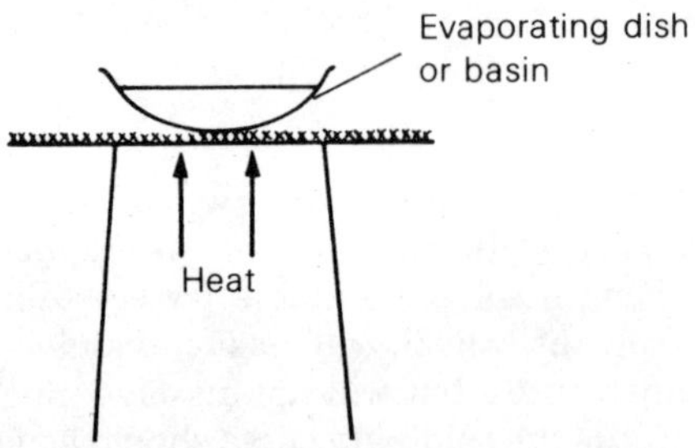

Fig. 7 Evaporation

Heat speeds up evaporation – pools of water "dry up" more quickly when the sun shines – and salt is soon left in the evaporating dish. If the liquid starts to spurt out near completion of evaporation, remove the burner and carefully use a small flame from time to time.

Distillation

In the evaporation of water from salt solution we can obtain the salt as residue but the unfortunate drawback is that the water is "lost" to the air. Suppose we want to avoid this and recover the water. How could this be done? You will have noticed that during cold weather the inside of room windows "mist up". This is because the water vapour produced by respiration is cooled by the cold glass of the windows and is changed back to liquid water. This change is called *condensation*; it is the very opposite of evaporation.

$$\text{WATER} \underset{\text{Condensation}}{\overset{\text{Evaporation}}{\rightleftharpoons}} \text{WATER VAPOUR}$$

The apparatus (Fig. 8) illustrates the method used to evaporate a liquid and then to condense the vapours. The process is called distillation.

The Liebig condenser

This consists of a glass tube surrounded by an outer jacket through which water can be passed. As the water vapour enters the glass tube it is cooled by the outer water jacket and is changed back to water. The water is thus condensed and can be collected.

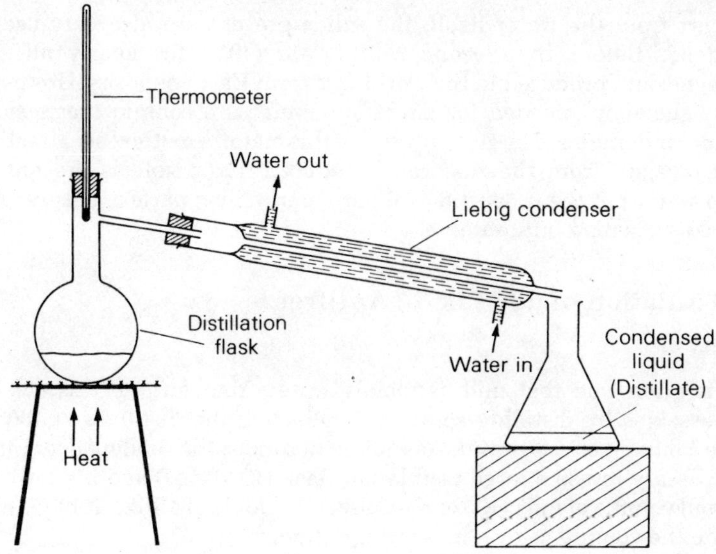

Fig. 8 Distillation

The water problem

Water is the most common of all liquids and probably the most useful. Naturally occurring forms such as sea-water, rain-water and lake-water are never pure. Consumption increases annually, millions of tons are used each day in industry, so in this respect there is no wonder that there is a water problem.

The answer to heavy demands might well be in the sea which covers most of the Earth's surface.

Sea-water varies in composition, but is, in fact, a vast store house of substances apart from water. In the English Channel there is about 3·5 per cent of solids dissolved in the water, in the Red Sea about 4·1 per cent while in the Baltic it can be as low as 0·5 per cent. Sodium chloride, common salt, is the most abundant of the solids present. From your knowledge of separation so far, you can easily understand that distillation of sea-water would give pure water and leave the solids in the distillation vessel. Of course this has been realised for a long time but the difficulty has been in carrying out this operation economically. To satisfy the heavy demands of industry, a great deal of fuel would be needed to carry out the distillation, to mention but one snag. However, using the Sun as the source of heat, purification of sea-water has successfully been carried out in Israel in recent years.

Uses of sea-water

Apart from the water itself, the solids present are also very useful indeed. Before the Second World War (1939–45) nearly all the magnesium produced in the world was from Earth deposits. However, as light alloys needed for aircraft manufacture contain magnesium and war demanded large quantities of this metal, a method of extracting magnesium from the sea was introduced. Even gold is present in sea-water but the proportion – 6 parts per million parts of sea-water – makes isolation unprofitable.

Distillation of (a) Milk (b) Antifreeze

Milk

It is often said that milk is mainly water. You can find this out for yourselves by distilling some and collecting the distillate at 100°C, the boiling point of water. As milk is notorious for "boiling over", it is necessary to use a large distillation flask (1000 cm^3) and a relatively small volume of milk (100 cm^3). Collect the liquid distilling at 100°C and note the volume using a measuring cylinder.

Is milk mainly water?

N.B. When the volume left in the flask becomes small it may tend to burn. If this happens, then stop heating the flask.

Repeat the experiment using ink instead of water. Note your results. Can you explain them?

Antifreeze

This varies in colour according to the make. Shell or B.P. grades are blue and suitable for this experiment. Pour antifreeze into a distillation flask so that it is about one-third full and fix a 250°C thermometer into the cork. Add a few pieces of porous pot to the antifreeze for even boiling. Heat the flask and contents until the thermometer shows a temperature of about 120°C. At this point stop heating the flask and either let the water run out of the condenser or replace the condenser by a glass tube. The "air condenser" is sufficient to condense vapours above 120°C and in any case the glass condenser would crack if subjected to cold water and temperatures of over 120°C at the same time.

Now, heat the liquid in the flask again and note the temperature at which most of the liquid distils. This temperature is the boiling point of the liquid. Record your results.

Boiling point	
Colour of distillate	

The liquid is called ethanediol.

Fractional distillation

If we have a mixture of two liquids A and B and A is more volatile than B, it is usually possible to separate the two by fractional distillation.

When a mixture of A and B is boiled, one might expect the vapour to be pure A, since it is more volatile. This is not so, because some B is present too, but in a much smaller proportion. If this vapour is cooled and boiled again, the vapour now will have an even greater proportion of A. Now if this process is repeated several times the vapour will eventually be all A.

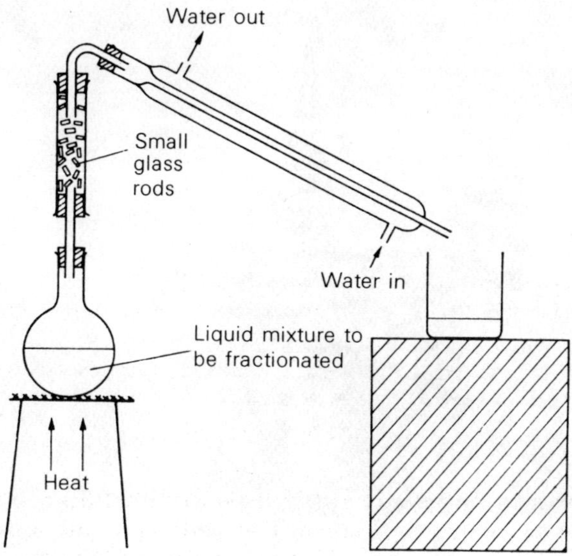

Fig. 9 Fractional distillation in the laboratory

In the laboratory we can use the apparatus shown (Fig. 9) in which the tube of small glass rods acts as the fractionating column. The liquid is evaporated and a large amount condenses on the small glass rods and runs back into the flask, to be evaporated again. As the column gets warm the vapour of the more volatile liquid rises to the top and distils over.

Petroleum or crude oil

This is one of the world's most valuable liquid mixtures. It varies widely in appearance ranging from an almost colourless, volatile fluid to a thick black oil. Generally, it looks rather like thin treacle. The world's four major regions, as regards proven reserves, are the Middle East, the U.S.A., the Caribbean and the U.S.S.R. Crude oil has been obtained by hand for centuries but it was not until 1848 that oil was actually struck during borings for salt water in Pennsylvania, U.S.A. Eleven years later "Colonel" Drake became the first man to drill a well specifically to find oil, and to be successful. The deposits that are present in the rock, deep under the North Sea, make Britain an important oil producer and exporter (Plate 1).

Plate 1 U.K. North Sea oilfield in production

Laboratory distillation of petroleum (demonstration)

Petroleum can be soaked in Rocksil and heated in a test tube (Fig. 10). The tube can be heated and the distillate collected in four tubes:

1st tube	up to 75°C	3rd tube	125–175°C
2nd tube	75–125°C	4th tube	175–225°C

The fact that the petroleum distils over a wide range of temperature shows that it is a mixture of liquids. If it were a pure liquid it would distil at one particular temperature, e.g. water 100°C; ethane-1,2-diol (ethylene glycol) 197°C. You can test the four samples collected:

(i) Note the colour of each.
(ii) Smell each one.
(iii) See which flows easiest, which with most difficulty.
(iv) Put a lighted splint to each.

Fig. 10 Simple distillation of oil

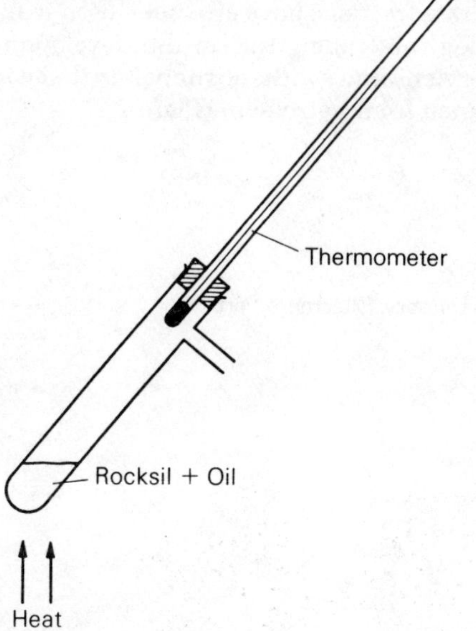

The petroleum industry

Fractional distillation of petroleum is carried out on a very large scale in industry. You may have seen the very tall towers at an oil refinery in which the fractional distillation is carried out. While it would not be convenient at this introductory stage to give the full story, the industry is so important and continues to expand so much that a little about the products of separation of the mixture should be known.

Table of the fractions

		Approximate temperature of distillation
Petroleum	Petrol (gasoline)	20–200°C
	Kerosene	200–300°C
	Gas oil and diesel oil	280–360°C
	Lubricating oil	Above 360°C
	Fuel oil	Above 360°C
	Bitumen	Left behind at about 400°C

Before the 1920s petroleum was almost entirely used as the source of petrol, kerosene, diesel oil and lubricating oil. During the last 50 years or so, these separate fractions have also been used to make chemicals, the U.S.A. playing the leading role in the development. Polythene, PVC, perspex, terylene and synthetic rubber are but a few of the many substances obtained from petroleum (Plate 2).

Plate 2 Stanlow oil refinery, Ellesmere Port

How do we know whether a substance is pure?

Most pure substances melt at a fixed temperature and boil at another fixed temperature. These temperatures are really only fixed when the pressure on them is constant.

Water, when pure, will boil at 100°C at normal atmospheric pressure. Ethane-1,2-diol will boil at 197°C.

Ice, when pure, will melt at 0°C, naphthalene will melt at 80°C.

How to find a melting point

Put a little of the solid in a small tube and attach it to a thermometer with a rubber band. Place this arrangement in a liquid which will boil at a temperature considerably higher than the melting point of the solid. The mouth of the tube must be above the liquid. Heat the liquid and note the temperature at which the solid melts (Fig. 11).

If it melts sharply at a particular temperature the solid is pure and the temperature is called the melting point. If impurities are present the solid will melt at a temperature lower than the melting point of the pure solid and the melting will take place gradually over a range of several degrees.

Fig. 11 Taking a melting point

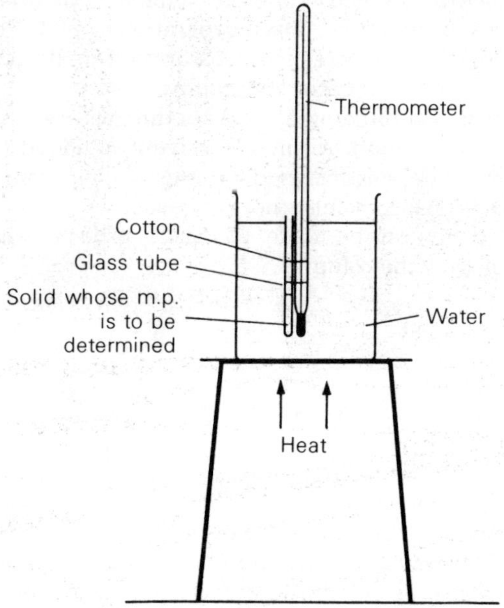

3. ADSORPTION

When litmus solution is boiled for a few minutes in the presence of animal charcoal and then filtered, the filtrate is colourless. The animal charcoal has removed the dye from the litmus and holds it at the *surface* of the charcoal. The process is called *adsorption*. Charcoal will adsorb other substances and was at one time used in gas masks to adsorb poisonous gases. Today, adsorption is used to separate mixtures.

Chromatography

The story of chromatography began about 60 years ago when the Polish botanist, Tswett, carried out experiments on the pigments of green leaves. Tswett dissolved the coloured substances from green leaves in petroleum ether and poured the solution over finely powdered calcium carbonate in a vertical glass tube (Fig. 12). As the solution seeped through the colourless calcium carbonate he noticed that the top of the calcium carbonate was yellow while lower down it was green. There were, in fact, two distinct bands or zones. The yellow band was caused by a substance called carotene; the green band by chlorophyll. These two substances were separated because calcium carbonate adsorbs one more easily than the other. This kind of separation is called *chromatography* (Greek *chroma*, colour) because in the early years the method was used only for coloured mixtures. In modern times the method has been extended to colourless mixtures, the substances from which, after separation, can be detected by treatment with a suitable reagent. Adsorption analysis would be a more suitable name than chromatography.

In the process of chromatography, after the mixture has been added at the top of the column, a suitable solvent is added to "develop" the separation. The solvent used for the original solution is tried first but if this is unsuccessful, others are used.

A list of solvents can be made, giving the order in which they are tried for developing the column.

1. Petroleum ether
2. Ether
3. Propanone
4. Benzene
5. Alcohol
6. Water

After separation, the column can be pushed out and cut up to separate the bands. A suitable solvent can then be used to dissolve each substance, and after filtration of the solution from the column material the solute can be crystallised to purify it.

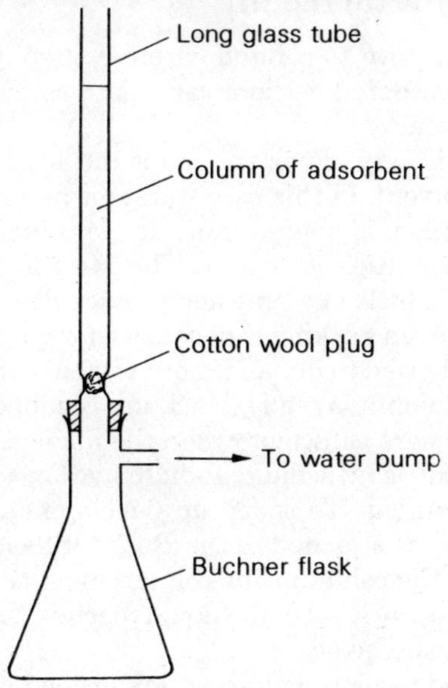

Fig. 12 Column chromatography

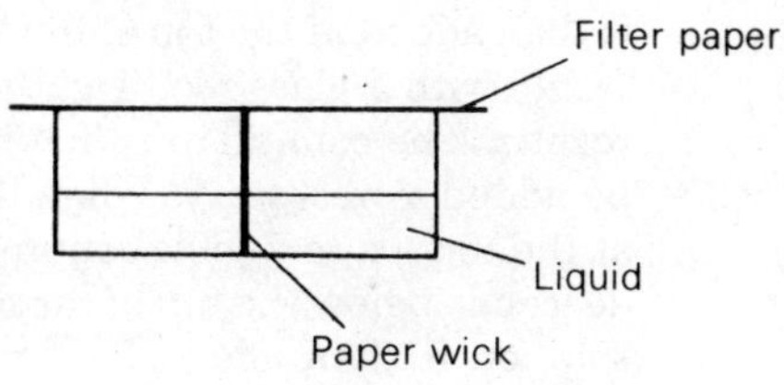

Fig. 13 Paper chromatography (simple wick method)

Terms used

Adsorbent. The material used for the column, e.g. calcium carbonate (powdered form), aluminium oxide.

Chromatogram. The column with its adsorbed substances after separation.

Development. This involves the separation of the mixture on the column by allowing a suitable liquid to pass through.

Paper chromatography

Paper with good adsorbent properties can be used to separate mixtures. A very simple method which can be used to illustrate paper chromatography is as follows. A filter paper is supported on an evaporating dish or a Petri dish which contains the solvent mixture. Another filter paper is folded to form a wick and this can be fixed upright so that it touches the underside of the filter paper and also dips into the liquid (Fig. 13). The liquid is drawn up by the wick and transferred to the centre of the filter paper. As the liquid spreads out from the centre, the different substances in the mixtures are adsorbed at different rates and instead of bands, as in column chromatography, concentric rings are produced on the filter paper.

Experiments with (a) black ink, (b) red ink

These two inks can be subjected in turn to column chromatography using either aluminium oxide or powdered calcium carbonate as the adsorbent and water for development.

The preparation of the column is very important and is the key to successful chromatograms. The solvent, in this case water, or better still, dilute sodium hydroxide solution, is poured into the glass tube which is plugged at the bottom with a little cotton wool. The adsorbent is then added at the top of the tube a little at a time and pressed down firmly with a glass rod. Tight and even packing is necessary for good results. The column must not be allowed to dry and more solvent can be added if necessary. When the column is ready, black ink is added at the top – a few cubic centimetres are sufficient. when the ink level descends below the top of the column, a little dilute sodium hydroxide solution is then added for development. To speed up development of the chromatogram a water pump is attached to the Buchner flask (Fig. 12) to draw the solution down the column. The column must not be allowed to become dry. As soon as the level of the liquid reaches the top of the column, more solvent should be added.

The experiment can be repeated using red or green ink instead of black ink; beetroot juice can be used with alcohol as solvent, or the pigments in grass can be separated using the solvent propanone.

To obtain a paper chromatograph of black ink (alternative methods)

1. Place a filter paper between two sheets of glass, the upper one of which has a small hole in the middle. Then place the sheets of glass on the bench. Using a dropping tube put a drop of black ink into the hole and then place weights on the upper sheet of glass. Add drops of dilute sodium hydroxide solution into the hole to develop the chromatogram.

2. Place a spot of black ink on a strip of filter paper which is about 25 cm long and 5 cm wide. The ink spot should be about 4 cm from one end. Then pin the paper to a wooden splint and place the splint across the top of a gas jar which contains dilute sodium hydroxide solution to a depth of about 2 cm. The paper is made to hang lengthwise in the jar so that the ink spot is *just above* the liquid in the gas jar. As the liquid rises up the paper the mixture of colours in the ink is separated.

Uses of chromatography

Chromatography has many applications; one of the most important involves the testing of the colouring matter added to some foods and drinks, such as tinned peas, sausages, sweets and soft drinks, to find out whether it is medically safe.

Manufacturers of foods and drinks are only allowed to use certain colouring materials. There are about two dozen permitted substances at present, but there has been some discussion as to whether this number should be reduced. There are several substances with different shades of red, several blues and yellows. The green colour in some foods is usually produced by a mixture of permitted blue and yellow materials. If you ever have the chance to visit a public health laboratory you may find chemists carrying out chromatography tests on foods and drinks which have been made in that particular area. The colouring materials used in the foods and drinks may be tested using the method given on page 21. The test depends on comparing the rate at which the colouring substance is adsorbed on the adsorbent paper with the rates at which permitted substances are adsorbed.

The purification of sugar

After collection from the fields the sugar-cane is crushed by heavy rollers to squeeze out the sweet sugar juices. Then, after the removal of impurities using milk of lime, the clear juice is concentrated to obtain crystals of raw sugar which are separated from the remaining syrup by centrifuging. The crystals are then dried and purified by treatment with "bone char" – a form of carbon made by heating bones in the absence of air. The "bone char" *adsorbs* the impurities from the raw sugar and leaves the pure cane-sugar.

CHEMICAL AND PHYSICAL CHANGES

So far in this chapter we have studied changes such as evaporation, distillation and adsorption. Let us look for a moment at the changes which water can undergo. If we cool water sufficiently it will freeze and become solid ice, but on raising the temperature again, the ice melts and water is re-formed. We can go on heating the water and, if we heat it sufficiently, it will boil and will all eventually be turned to vapour. Earlier in the chapter we have seen that if this is done in a distillation apparatus the vapour will condense in the condenser and water will be re-formed. We can represent these changes thus:

$$\text{ICE} \underset{\text{Cool}}{\overset{\text{Heat}}{\rightleftarrows}} \text{WATER} \underset{\text{Cool}}{\overset{\text{Heat}}{\rightleftarrows}} \text{WATER VAPOUR}$$

The changes are reversible and the substance in each case is a form of water, that is, no *new* substance is produced. Such changes are known as physical changes. Evaporation, condensation and adsorption are all, in fact, physical changes.

Let us now consider the burning of coal. When this takes place, smoke is given off, ashes remain and much heat is produced during the change. However, we cannot collect the ashes and smoke and heat them or cool them to make the coal again, that is, the change is not reversible. This is a good example of a chemical change where there is a new substance formed and the change is not reversible. Very often much heat is given out or taken in during a chemical change.

Thus we see that there are two types of changes and the table set out below shows how the type can be recognised.

Chemical changes	Physical changes
1. New substance formed.	1. No new substance formed.
2. Usually not reversible.	2. Reversible.
3. Heat given out or taken in.	3. Very little, if any, heat exchange.

In this chapter we have dealt mainly with physical changes. Chemistry is mainly concerned with chemical changes and we shall study a great number of this type in the following chapters.

THINGS TO REMEMBER

- The filtrate is the liquid which passes through the filter paper during filtration.
- Evaporation is the change from liquid to vapour without necessarily reaching the boiling point.
- Distillation is the process of converting a liquid into vapour, condensing the vapour, and collecting the condensed liquid.
- The distillate is the liquid formed by condensing the vapour during distillation.
- A pure substance will melt and boil at certain temperatures provided there is no decomposition.
- Melting points and boiling points are sometimes used to identify unknown, pure substances.
- Solvents dissolve solutes to give solutions.
- Paper chromatography is a means of separating a mixture by allowing the mixture to spread on paper.
- Fractional distillation is the separation of a mixture of liquids which have different boiling points, by collecting separately "fractions" boiling at different temperatures.

QUESTIONS TO TRY

1. What kind of process would you use:
   (a) to obtain sugar from a sugar solution;
   (b) to obtain sand from a mixture of salt and sand;
   (c) to obtain water from soap solution;
   (d) to separate the pigments in grass;
   (e) to separate the fractions which make up petroleum?

2. Describe what you would do to separate the first-named substance from each of the following pairs:
   (a) sugar and chalk;
   (b) salt dissolved in water;
   (c) ethanediol from antifreeze.

3. If you were given two samples of water, one of which has been distilled and the other containing a small quantity of solids, describe how you would find out by the quickest possible method which of the two samples was pure water?

4. Gunpowder is a mixture of potassium nitrate, sulphur and charcoal, how would you separate this mixture knowing that:
   (a) potassium nitrate is soluble in water but not tetrachloromethane;
   (b) sulphur is soluble in tetrachloromethane but not in water;
   (c) charcoal is soluble in neither water nor tetrachloromethane.

5. Classify the following as physical or chemical changes:
   (a) Melting of ice
   (b) Souring of milk
   (c) Rusting
   (d) Dissolving sugar in water
   (e) Separating the pigments of grass by chromatography
   (f) Burning wood.

   Give your reasons.

6. Draw the solubility curve for potash alum from the following results:

Solubility (g per 100 g water)	4	9	15	22	31	44	67	91	135	209	358
Temp. (°C)	0	10	20	30	40	50	60	70	80	90	100

   What would be the solubility of potash alum at 75°C? Do you think water would be a good solvent for crystallising potash alum? Give your reasons.

7. Draw the solubility curve for ammonium chloride from:

Solubility (g per 100 g water)	29	33	37	41	46	50	55	60
Temp. (°C)	0	10	20	30	40	50	60	70

   What is the solubility at 45°C? Would crystallisation of ammonium chloride be easy using water as the solvent?

8. If you were given an aqueous solution of salt and phenol, how could you obtain a sample of the latter solid knowing that it is much more soluble in ether than in water. Note that salt does not dissolve in ether, and water and ether do not mix very well. Use the apparatus in Fig. 14 and any other you have met in Chapter 2.

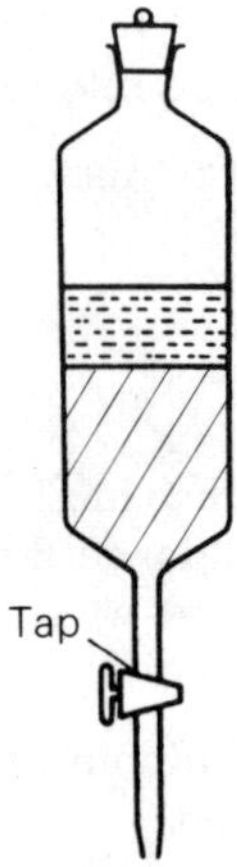

Fig. 14 The separating funnel

9. A firework was found to contain a mixture of potassium nitrate, carbon, sulphur, iron and magnesium. Of these, only potassium nitrate is soluble in water. How would you make some potassium nitrate crystals from the mixture?

10. Give *three* different physical changes which take place in the process of obtaining cane-sugar from sugar-cane.

3 Gases of the Air

The air, or atmosphere as it is sometimes called, is a mixture of gases which forms a blanket round the Earth. The total weight of the air is about 5000 million tonnes. The pull of the Earth, or its force of gravity, holds the atmosphere close to it, but as this force decreases as the distance from the centre of the Earth increases, the atmosphere becomes "thinner" or less dense the farther it is from the Earth. This means there is less oxygen to breathe when we get high up above the Earth and the scarcity of oxygen makes breathing difficult. This is why Hillary and Tensing had to use oxygen cylinders to make their breathing less difficult when they climbed Everest in 1953, and why some athletes experienced difficulty in the 1968 Olympic Games in Mexico City, altitude in excess of 20 000 m. Space exploration outside the Earth's atmosphere, including the momentous Moon landing by Neil Armstrong and Edwin Aldrin on 21 July 1969, would not have been possible without oxygen supplies for breathing and rocket propulsion.

Some tests on air

A. *The iced water experiment*

Put a few ice-cubes in a beaker of water and watch what happens. You will see that the beaker becomes "misted-up". This is because water vapour, which is present in the air, is condensed on the cold surface of the beaker and forms small drops of water thus giving a misty effect. To make sure that the water comes from the air and not from that inside the beaker, do the same test again this time using a conical flask which you can cork tightly.

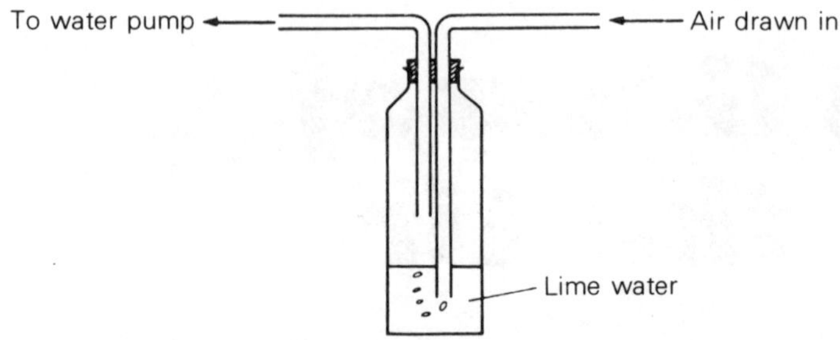

Fig. 15 Drawing air through lime water

B. *Lime water test*

Draw air through some lime water in the apparatus shown (Fig. 15) for about half an hour. Again watch what happens. Does the air coming out of the apparatus look any different from that going in? Try passing the air which comes out through another bottle of lime water and watch what happens. What would you conclude from the results of your experiment?

C. *The part of the air that helps things to burn*

When copper is heated in air it goes black. This is because it takes in the part of the air that is used up when things burn.

If some copper wire is heated in a silica glass tube and held between two gas syringes (Fig. 16), then the volume of the gas that is taken in can be measured, and if 100 cm^3 of air is taken into one syringe, with the other syringe empty, then the percentage of the gas that is needed when things burn can be measured.

Example

Volume of air in the syringes before the copper is heated $= 100\ cm^3$

Volume of air in the syringes after heating the copper, and after the apparatus has cooled $= 79\ cm^3$

$\therefore$ the percentage of the gas taken up $= \frac{(100-79)}{100} \times 100$

$= 21\%$

From the results we can see that only 21 cm^3 of the 100 cm^3 is used in "burning" the copper. This is about one-fifth of the air, so four-fifths of the air will not allow things to burn.

We have therefore learnt something about the composition of air. First, it contains water vapour which can be condensed to water by

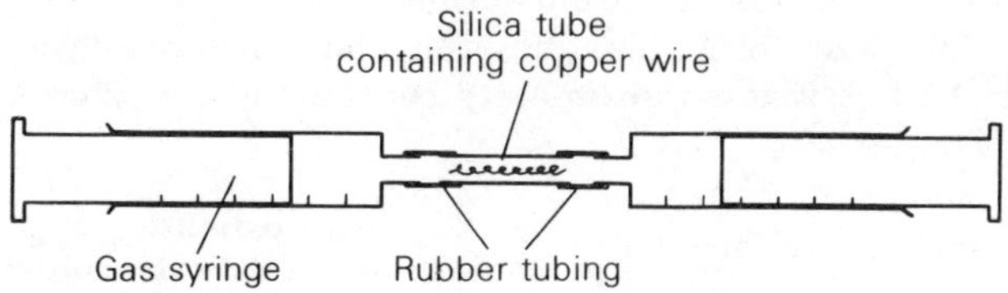

Fig. 16 Testing the volume of air used to burn some copper wire

bringing air in contact with a cold surface; second, it contains a gas which turns lime water milky. We shall learn more about this gas later; it is called carbon dioxide. Third, only one-fifth of the atmosphere is used up when a substance is burnt, and substances will not burn in the remaining four-fifths that is left.

If an animal was placed in the gas that remained after the copper had been heated then the animal would die. Therefore, the gas which was removed by the copper is the gas we need to live, i.e. oxygen. The gas which is left and which does not support combustion is mainly nitrogen.

Composition of air

Air is certainly a mixture. As we have seen, it contains oxygen, water vapour and carbon dioxide and until about 80 years ago it was believed that the remainder was nitrogen. In 1894, Lord Rayleigh carried out experiments which showed that the "nitrogen", which does not readily combine with other elements – we have seen it is left when copper and other substances 'burn' in air – contained a very small percentage of gases which were even less ready to combine or, as we say, even more inert, than the nitrogen itself. In a brilliant piece of work these gases were separated and were called the "inert" gases as it was thought that they did not combine to form any compounds at all. Quite recently it has been found that they do form a few rather complex compounds so the name has had to be altered and we now call them the "noble" gases. The gases are neon, argon, krypton, xenon and helium, and altogether total about 1 per cent of the atmosphere. Helium is interesting because it was detected in the Sun in 1868, well before it was found on Earth. It is a very light gas, only one other substance, hydrogen, is lighter, and is used for filling meteorological balloons. It is better for this purpose than hydrogen because hydrogen will burn and catches fire if it is near a naked light whereas helium does not burn – it is non-combustible. Neon is used in strip lighting to fill the tubes for fluorescent purposes and electric light bulbs are sometimes filled with argon since, as it is an inactive gas, it prevents the filament from burning away.

The composition of the air varies from place to place and also varies with height but it is approximately constant and is shown in the following table.

Substance	Approximate percentage composition by volume
Nitrogen	78
Oxygen	21
Water vapour	up to 4
Rare gases	1
Carbon dioxide	0.03
Dust, dirt, industrial gases	small amounts

As you see from the table, the main constituent is nitrogen but probably the most reactive gas is oxygen, since it allows substances to burn in it. We are going to learn more about oxygen and also about carbon dioxide, the gas which is formed when coal and many other animal and vegetable products are burned, and which turns lime water milky.

THE STORY OF OXYGEN

It is very easy in this modern age to accept, without much appreciation, valuable work which was done quite a long time ago. Two men stand out for their pioneer work on air and oxygen. One was English, Joseph Priestley; the other was French, Antoine Lavoisier.

The work of Priestley

Joseph Priestley was born in Yorkshire in 1733 and became a Nonconformist minister as his family wished. He was very interested in chemistry and became an expert in preparing and handling gases, being one of the first people to use a pneumatic trough, a special piece of apparatus, for gas collection. There was no Bunsen burner in those days, Robert Bunsen was not born until 1811 and did not introduce his burner until 1855, so it was not an easy matter to heat substances without destroying them. Priestley was presented with a large lens by some admirers in 1774 and he used it to direct the sunlight on to substances and so raise their temperature. One of these substances was the red calx of mercury, now called mercury oxide. He found that when this substance was heated, it gave off a colourless gas which he collected and he called this gas "dephlogisticated air"; we now call it oxygen.

$$\text{Red calx of mercury} \xrightarrow{\text{heat}} \text{mercury} + \text{oxygen}$$

Priestley found this gas very interesting as substances burned better in it than they did in air; for instance a candle burned much more brightly in the gas. In 1775 he visited Paris and met Lavoisier, the leading French chemist of that day, who found Priestley's work very stimulating and himself carried out what has now become a classical experiment.

Fig. 17 Lavoisier's apparatus

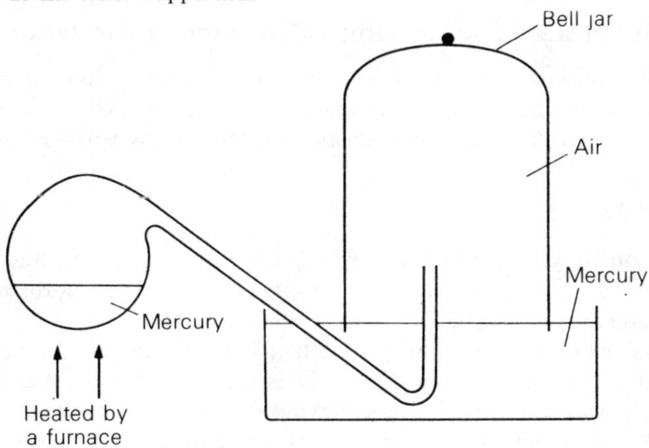

The work of Lavoisier

After seeing Priestley, Lavoisier was inspired to use the apparatus shown (Fig. 17). He heated mercury in the flask and noticed the appearance of red particles on the surface of the mercury. The level of the mercury in the bell jar began to rise at the same time, showing that some of the air in the bell jar was being removed. Lavoisier concluded that some part of the air was used in the formation of the red calx of mercury. During 12 days of almost continuous heating the amount of red calx increased and the level of the mercury in the bell jar rose higher. At the end of this time no further changes took place.

He found that the gas remaining in the bell jar occupied four-fifths of the original volume of gas and it extinguished a lighted splint and suffocated animals. He called the gas azote; the French still do. We call it nitrogen. (Today, of course, we know that although nitrogen occupies the bulk of the four-fifths remaining, traces of other gases are present.)

Lavoisier knew from Priestley's work that red calx of mercury decomposed when heated to give oxygen. He therefore heated the red calx he had obtained to a higher temperature and collected the evolved gas. The volume of this gas was found to be equal to the volume of gas used up in the first part of his experiment and, what is more, it was found to be an excellent supporter of combustion. Thus, the modern theory of combustion was born – oxygen, one fifth of the air by volume, supports combustion and is used in the process. The rest of the air does not support combustion and remains unchanged.

Lavoisier was a victim of the French Revolution and was guillotined because of his aristocratic connections. A sad and an unfitting end to a great man.

Preparation and collection of oxygen in the laboratory

Not only mercury oxide gives off oxygen on heating. Other substances capable of producing oxygen on heating include: potassium chlorate, potassium nitrate, potassium permanganate and hydrogen peroxide.

Catalysts

A solution of hydrogen peroxide slowly decomposes on standing. Light and heat both help to bring this about. Pure hydrogen peroxide has been used as a rocket fuel.

Take a boiling tube and put into it 5 to 10 cm^3 of "20 volume" hydrogen peroxide solution. Then place into it one spatula measure of manganese dioxide, and lower a glowing splint into the boiling tube.

The manganese dioxide undergoes no change in mass during the reaction. It is called a *catalyst*.

A catalyst is a substance which speeds up a chemical change, but remains unchanged in mass at the end of the reaction.

Oxygen from hydrogen peroxide

Although oxygen may be obtained from hydrogen peroxide by the action of heat, when a catalyst such as manganese dioxide is present, oxygen is liberated at room temperature.

Allow diluted hydrogen peroxide solution to fall on to manganese dioxide in a flask fitted up as shown (Fig. 18). As the reaction proceeds at room temperature the flow of oxygen can be controlled to some extent by the careful use of the tap funnel. Collect the evolved gas over water in gas jars but discard the first jar as it will chiefly contain air displaced from the apparatus. Collect six jars in all.

$$\text{Hydrogen peroxide} \xrightarrow{\text{manganese dioxide catalyst}} \text{water} + \text{oxygen}$$

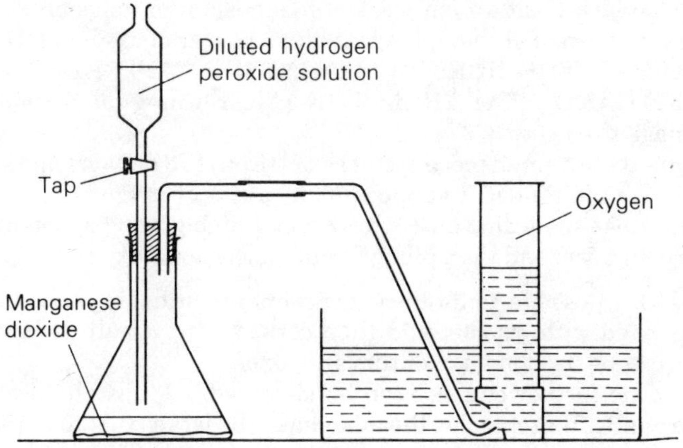

Fig. 18 Oxygen from hydrogen peroxide

Tests on the gas

Watch carefully what happens when you carry out these experiments. Make sure that you wear eye shields for all these experiments.

1. Remove the lid of one jar of gas, and lower a glowing splint into it.
2. *Carbon.* Heat a small piece of charcoal on a deflagrating spoon, and lower the hot charcoal on the spoon into a jar of oxygen (Fig. 19).

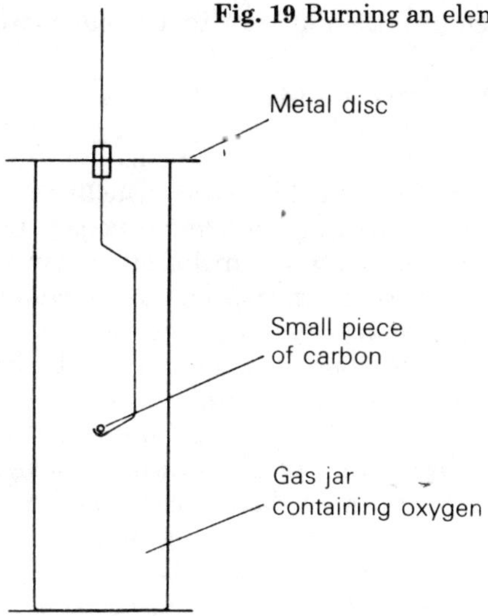

Fig. 19 Burning an element in oxygen

3. *Sulphur.* Heat some sulphur on a spoon until it just catches fire and burns slowly. Lower the spoon and sulphur into a jar of oxygen.
4. *Magnesium.* Heat a 1 cm piece of magnesium in tongs until it just starts to burn, and then plunge it into a gas jar of oxygen. DO NOT LOOK AT THE BURNING MAGNESIUM AND KEEP YOUR FREE HAND AWAY FROM THE JAR. Burning magnesium can damage the eyes.
5. *Iron.* Attach a small piece of steel wool (iron) to a deflagrating spoon, and warm it. Plunge it at once into a gas jar of oxygen.
6. *Zinc.* Heat some zinc dust strongly in a deflagrating spoon until it starts to burn, and then plunge it into a gas jar of oxygen.

It is also possible to do these experiments in boiling tubes which can be filled with oxygen, and then corked. Very small deflagrating spoons can be used instead of the large ones.

When no further changes can be observed, add distilled water to each jar, and after replacing the lids, shake the jars so that the products of the reactions come into contact with the water. Then add a little universal indicator, and record the pH of each liquid. Enter your results in a table in your book.

Your teacher might also burn some dangerous substances in oxygen. These could include small amounts of red phosphorus, sodium and calcium.

Look at the completed table and you will see that by using universal indicator you can divide the substances which burn in oxygen to give oxides soluble in water into two classes.

1. Those which turn the indicator to a colour representing a pH of less than 7.

2. Those which turn the indicator to a colour representing a pH of more than 7.

The first group therefore produces *acidic* oxides and the second group *alkaline or basic* oxides.

The substances which produce acidic oxides are carbon, sulphur and phosphorus and these substances are usually without lustre, break up to a powder on hammering, and do not behave like metals in any way. We call them non-metals. On the other hand, the substances which give alkaline or basic oxides, sodium, calcium and magnesium, have a silvery metallic look (sodium and calcium should be freshly cut to see this) and on hammering do not usually break, but flatten out. These, then, are metals and when they are burnt or heated in oxygen they give oxides which, if soluble in water (sodium and potassium oxides are very soluble) turn the universal indicator to a colour showing a pH of more than 7. Not many metallic oxides are soluble in water; calcium oxide is fairly soluble and magnesium oxide very slightly soluble. Iron oxide and zinc oxide are not soluble, but we shall see later that they behave as basic oxides, as do the oxides of all the metals.

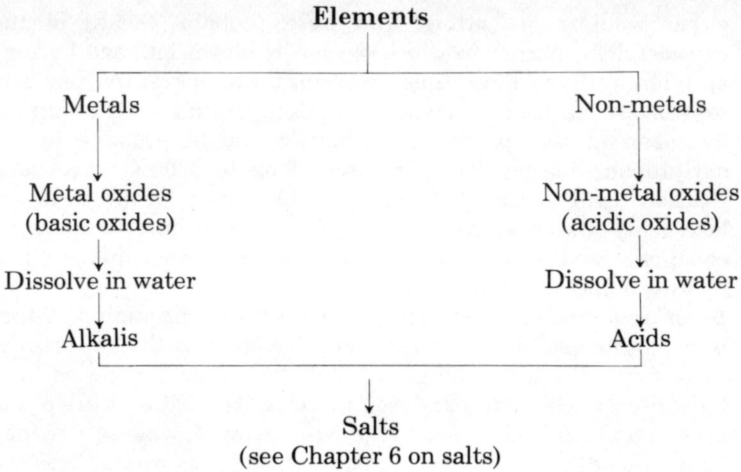

When acidic oxides are placed in water, a chemical reaction occurs, and an *acid* is produced.

Sulphur trioxide + water $\longrightarrow$ sulphuric acid
Carbon dioxide + water $\longrightarrow$ carbonic acid
Phosphorus oxide + water $\longrightarrow$ phosphoric acid

Some basic oxides give a chemical reaction with water

Sodium oxide + water $\longrightarrow$ sodium hydroxide

Sodium hydroxide is an *alkali*.

SOME OTHER POINTS ABOUT OXYGEN

It is a colourless gas without smell. It is slightly soluble in water and this is a very important property since fish and animals which live in water are able to get the oxygen they need for respiration. As we have seen it is a very good supporter of combustion but the gas itself does not burn.

What is oxygen used for?

1. *Welding and cutting metals.* In these processes oxygen is used to burn gases such as hydrogen, acetylene and propane. Much heat is produced during the combustion and metals can be melted to allow

either welding or cutting. You have probably heard of the oxy-acetylene burner in which oxygen is blown into the burning acetylene (ethyne) flame thus increasing the temperature. Similarly oxygen can be used to increase the temperature of the burning hydrogen in the oxy-hydrogen burner and of propane in the oxy-propane burner. Temperatures of up to 3000°C have been reached using burners of this type. Demands for propane are increasing; its advantages over hydrogen and acetylene are its cheapness and it is not as dangerous as the highly inflammable hydrogen and acetylene.

2. *Breathing stimulant.* Oxygen is used as a breathing stimulant; that is it helps people who are having difficulty with breathing to breathe more naturally. It is used in hospitals for patients suffering from lung disorders or from paralysed muscles. The patients are put in oxygen tents. It is also used in mine rescue work, where there may be pockets of poisonous gas, and at high altitudes where there is a lack of air.
3. *Steelmaking.* Oxygen is made by the tonne to convert pig iron into steel. Plants producing several hundred tonnes of high purity oxygen per day have been built adjacent to some of the larger steelworks.

Oxygen by the tonne

The method described for the laboratory preparation would be far too expensive for industrial purposes. The air is the obvious source of oxygen and the aim has been to extract it as economically as possible. The industrial method therefore involves the separation of oxygen from the other gases of the air. This is done by liquefaction of the air and the gases are then separated by fractional distillation. Air is first freed from dust. Then carbon dioxide and water vapour are removed because they would freeze when the air was brought down to the low temperatures necessary to liquefy it and would so block the pipes. The air, freed from dust, carbon dioxide and water vapour, is now submitted to high pressure and cooled. If a gas is compressed greatly and then suddenly allowed to pass through a small opening into a large space, it expands enormously and so loses energy in the form of heat and becomes cooler. The air is subjected to this treatment many times; each time it becomes so much colder and eventually it becomes so cold that it liquefies. The pale blue liquid obtained consists of liquid nitrogen (and the rare gases) and liquid oxygen. The liquefied gases have different boiling points and, like any other liquids, can be separated by fractional distillation. The temperature of the mixture is allowed to rise and the nitrogen, which boils at −196°C, comes off first leaving the liquid oxygen. The liquid oxygen can then be transferred to bulk containers for delivery or allowed to evaporate and piped to where it is required. It is very pure.

Lime water test

Collect two jars of oxygen. Burn a small piece of carbon in one and a small piece of calcium in the other. The carbon will produce a colourless gas, carbon dioxide, and the calcium will burn to a white powder, calcium oxide. Dissolve this in water – it is not very soluble and after shaking for some time you may have to filter some of the solid off as you want to use the clear calcium hydroxide solution which we call lime water. Pour this clear solution into the jar of carbon dioxide and shake up. Note what happens – you ought to know what to expect. The milky appearance is due to the formation of insoluble calcium carbonate. This reaction is used to test for carbon dioxide (see later).

CARBON DIOXIDE

This gas is present in the air to the extent of 3 parts for every 10 000 parts of air by volume. It is the gas in "fizzy drinks" and is a very important substance as will be seen later. The most economical sources of the gas are the carbonates, which exist extensively in the Earth's crust, particularly calcium carbonate. Most carbonates give off carbon dioxide when heated, leaving the metal oxide.

$$\text{Carbonates} \longrightarrow \text{metal oxides} + \text{carbon dioxide}$$

Calcium carbonate – a very valuable raw material

You have probably been impressed at some time or other by the white cliffs of Dover, or by the stalactites and stalagmites seen in certain caves, or by the wonderful array of shells on the sea-shore. The substance of which all these consist has the chemical name calcium carbonate. This substance can exist in different forms, for example as chalk in the white cliffs of Dover, the cliffs of the Yorkshire coast and the Chiltern Hills, and as limestone in the Pennine Chain, the Cotswolds and the Mendips. Marble, coral and pearl are yet other forms of calcium carbonate.

Action of heat on limestone

Put a small piece of limestone on a piece of moist neutral litmus paper and make a note of any change. Then place the lump on an old wire gauze with a small hole in the middle so that the hole is covered (Fig. 20). A loop of nichrome wire to support the limestone may be used instead of an old wire gauze. Gently heat the limestone for about a

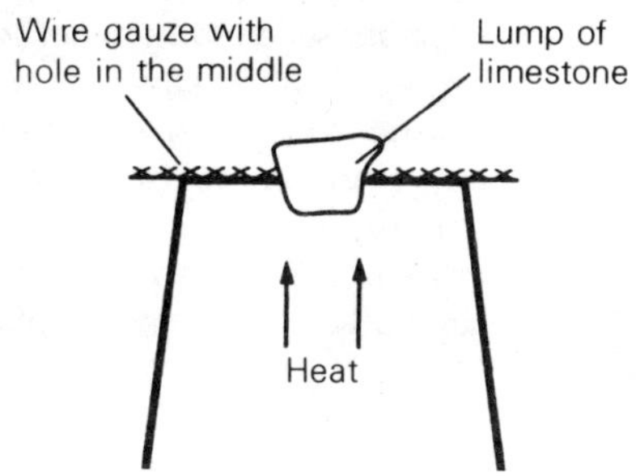

Fig. 20 Decomposition of limestone

minute and then heat it very strongly for about 15 min. Allow the product to cool, examine it and test it with a piece of moist neutral litmus paper. Then place the solid in a crucible, add about half a dozen drops of water, drop by drop, holding the crucible in the palm of the hand and note what happens.

Quicklime (calcium oxide) and slaked lime

Limestone, like all other forms of calcium carbonate, gives up carbon dioxide on heating leaving calcium oxide, commonly called quicklime.

$$\text{Calcium carbonate} \xrightarrow{\text{heat}} \underset{\text{(quicklime)}}{\text{calcium oxide}} + \text{carbon dioxide}$$

Calcium oxide is a basic oxide which on treatment with water produces much heat, causing some of the water to change to clouds of steam, disintegration of the solid and a *dry* white powder called calcium hydroxide or slaked lime.

$$\text{Calcium oxide} + \text{water} \longrightarrow \underset{\text{(slaked lime)}}{\text{calcium hydroxide}}$$

Lime water

Calcium hydroxide is slightly soluble in water forming an alkaline solution commonly called lime water.

Lime kilns

In industry, limestone is heated in lime kilns to produce quicklime on a large scale. The quicklime is converted to slaked lime by treatment with water and the slaked lime can be used in agriculture to counteract acidity in the soil. Much slaked lime is mixed with sand and water to make *mortar* which slowly hardens on standing as the water evaporates.

The action of acids on carbonates

Any carbonates can be used for this experiment. Familiar ones include washing soda (sodium carbonate), malachite (basic copper carbonate), egg shells and sea shells (calcium carbonate), etc.

Add diluted hydrochloric acid to washing soda in a test-tube and lead the gas into lime water. Note what happens. Repeat the experiment using the other carbonates in place of washing soda. If you have time repeat the tests with firstly dilute sulphuric acid and then dilute nitric acid in place of the dilute hydrochloric acid.

Test for carbonates

All carbonates liberate carbon dioxide when treated with dilute hydrochloric acid. The gas can be tested with lime water which is turned milky due to the formation of insoluble calcium carbonate.

Calcium hydroxide (lime water) + carbon dioxide ⟶ calcium carbonate (causes "milkiness") + water

The "fizzing" caused when acid attacks a carbonate is called effervescence. The corresponding *salt* is left in solution. For example:

Copper carbonate + hydrochloric acid ⟶ copper chloride (salt) + water + carbon dioxide

We shall learn more about salts in Chapter 6.

N.B. You may have noticed little, if any, reaction between calcium carbonate and dilute sulphuric acid. This is because the salt formed, calcium sulphate, does not dissolve in water to any appreciable extent and, therefore, as soon as it is formed in the reaction, forms a protective layer round the carbonate thus preventing further attack by the acid. When we wish to prepare some carbon dioxide we always use the method of adding a dilute acid, usually hydrochloric acid, to a carbonate, usually calcium carbonate though sodium carbonate would be equally suitable.

Hydrogencarbonates also will react with acids to give carbon dioxide.

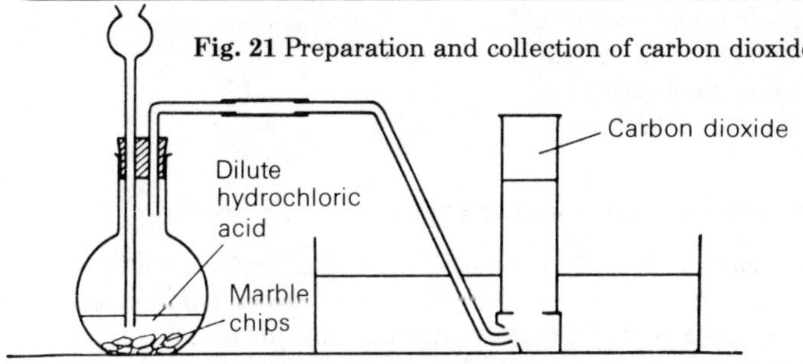

Fig. 21 Preparation and collection of carbon dioxide

An investigation into the properties of carbon dioxide

Put marble chips into a flat-bottomed flask and fit a two-holed stopper carrying a thistle funnel and a delivery tube. Pour dilute hydrochloric acid on the marble until the bottom of the thistle funnel stem is just covered. After allowing air to be displaced from the apparatus, collect the carbon dioxide gas over water (Fig. 21). Collect several gas jars and do the tests.

Reaction in the flask:

Calcium carbonate + hydrochloric acid ⟶ calcium chloride + water + carbon dioxide

Tests on the gas

1. *Lime water test.* After collecting about five gas jars pass the gas into a test-tube of lime water for several minutes until there is no further change. Then boil the mixture and note what happens.
2. Lower a lighted splint into the gas. What happens?
3. Light a small candle and place it in a beaker. Remove the lid of a gas jar and tilt the jar so that its mouth is just above the mouth of the beaker. Record what happens.
4. Place a beaker on the left-hand pan of a balance and add sufficient weights to the right-hand pan so that the pointer is at zero when the beam is raised. Raise the beam and "pour" the gas into the beaker. What conclusion would you draw from this?
5. Add neutral litmus solution to a gas jar and shake the contents well. Record what happens.
6. Wrap a 5 cm strip of magnesium ribbon round the ends of tongs, light one end and lower it into a jar of gas. Note particularly what happens and suggest an explanation.

After noting down all you saw you can write up the answers to the following questions:

(a) Does the gas burn in air?
(b) Does it support combustion?
(c) Is it denser than air?
(d) Is it acidic, alkaline or neutral?
(e) Is it colourless?
(f) Is it odourless?

Note on the lime water test

When carbon dioxide is passed into lime water a milky appearance results after a short time, but when more gas is led in, the "milkiness" disappears leaving a clear solution. The formation of the insoluble carbonate causes the "milky" appearance in the first instance.

$$\begin{array}{c} \text{Calcium} \\ \text{hydroxide} \\ \text{(lime water)} \end{array} + \begin{array}{c} \text{carbon} \\ \text{dioxide} \end{array} \longrightarrow \begin{array}{c} \text{calcium} \\ \text{carbonate} \\ \text{(insoluble)} \end{array} + \text{water}$$

Excess carbon dioxide reacts with calcium carbonate and water to form calcium hydrogencarbonate – a soluble compound. Therefore, the "milkiness" disappears.

$$\begin{array}{c} \text{Calcium} \\ \text{carbonate} \end{array} + \text{water} + \begin{array}{c} \text{carbon} \\ \text{dioxide} \end{array} \longrightarrow \begin{array}{c} \text{calcium} \\ \text{hydrogen} \\ \text{carbonate} \\ \text{(soluble)} \end{array}$$

When the clear solution of calcium hydrogencarbonate is boiled the milkiness reappears as the calcium hydrogencarbonate splits up to reform the calcium carbonate.

Calcium hydrogen carbonate → calcium carbonate (insoluble) + water + carbon dioxide

You will have noticed that the last reaction is the *reverse* of the one preceding it. We can represent the two reactions together:

Calcium carbonate + water + carbon dioxide $\rightleftharpoons$ calcium hydrogen carbonate

The sign $\rightleftharpoons$ shows that the reaction is reversible, i.e. it can be made to proceed in either direction.

Note on the reaction between carbon dioxide and magnesium

In this reaction, the magnesium will burn in the gas even though, normally, the gas does not support combustion. The products are magnesium oxide and carbon.

Magnesium + carbon dioxide → magnesium oxide (white solid) + carbon (black solid)

Stalactites and stalagmites

In some caverns, for instance, in the caves at Cheddar, pillars of limestone rising from the ground or hanging from the ceiling of the cave make a most impressive display. How are they formed? What is the chemistry behind their formation? If we know the chemistry of the lime water test then we should easily be able to understand the answers to these questions.

Rain water always has some carbon dioxide dissolved in it; it dissolves the gas as it passes through the air. When it falls on the ground and seeps through the soil into chalk or limestone regions, chemical reaction occurs and some of the calcium carbonate is converted to calcium hydrogencarbonate which is dissolved in the rain water.

Limestone (calcium carbonate) + rain-water (water + carbon dioxide) → calcium hydrogencarbonate

Some of the solution seeps through the ceilings of the caves and remains hanging from the ceiling in droplets until evaporation occurs. During the process of evaporation the calcium hydrogencarbonate (which does not exist except in solution) decomposes and a deposit of calcium carbonate remains on the ceiling. Similarly, some droplets of solutions fall to the floor of the cave, evaporate and leave a deposit of calcium carbonate (Fig. 22). These deposits grow extremely slowly over hundreds of years as more drops appear and more calcium carbonate forms.

Fig. 22 Formation of stalactites and stalagmites

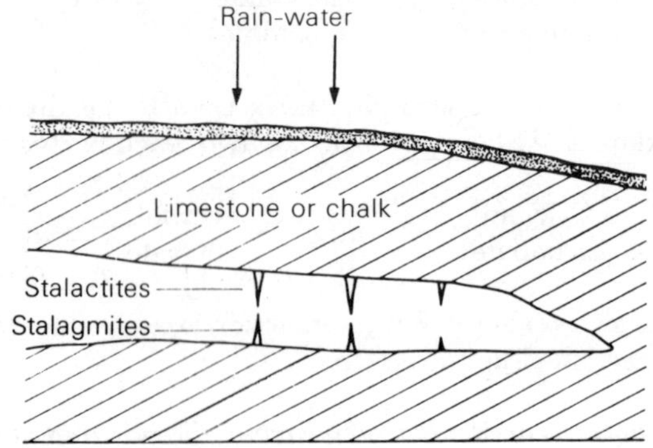

The formations on the ceiling are called stalactites. Those on the ground are called stalagmites. The Carlsbad Caverns (New Mexico), Cheddar (England) and Dordogne (France) are notable for their impressive formations.

What is carbon dioxide used for?

1. In mineral waters and soda water the gas is dissolved under high pressure and accounts for the sharp taste of the liquids.
2. Fire extinguishers.
   Water will not extinguish all fires. Oil fires, for example, are a problem because oil is less dense than water and does not mix with it. Therefore, when water is poured on to an oil fire, the oil simply floats on top of the water and continues to burn. Water will, in fact, spread the fire by providing a surface on which the oil can float. Carbon dioxide is effective as a fire extinguisher in this case because it is denser than air but not as dense as oil so that it can form a blanket over the fire and so prevent air reaching the flames.
3. Refrigeration – solid carbon dioxide (dry ice or drikold) is used here, and has a number of advantages over ice:
   (a) It does not melt to a liquid on heating but changes directly from solid to gas. This type of change is known as sublimation.
   (b) It is much colder than ice and a temperature of −79°C can be obtained.
   (c) It is not as dense as ice.
4. Medicine. In special cases it is used in *small quantities*, along with oxygen, as a stimulant to breathing.

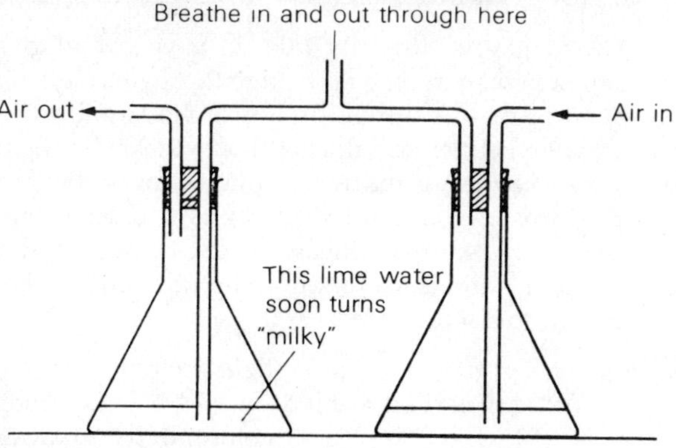

Fig. 23 Comparison between inhaled and exhaled air

Is inhaled air the same as exhaled air?

Arrange two conical flasks containing lime water so that air can be breathed in through one flask and out through the other (Fig. 23). What happens to the lime water in the flasks and what explanation can you give?

Oxidation of carbon in the body

Our food is made up of compounds containing chiefly carbon, hydrogen and oxygen. You will learn a great deal about the different kinds of food in your biology lesson. The air we breathe provides oxygen which combines with the haemoglobin — the red colouring matter of the blood. This "oxygenated blood" is then pumped round the body and oxidises the products of digestion producing carbon dioxide, water and energy. Does this fit in with your observations in the last experiment?

Fig. 24 Oxygen from pond weed

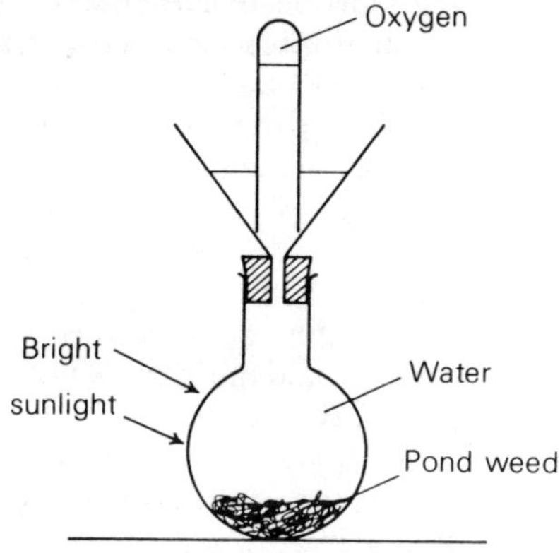

What is the gas which comes from green plants in sunlight?

Expose a piece of pond weed to bright sunlight for a few days in the apparatus shown (Fig. 24). Collect the gas and see if you can identify it.

Photosynthesis

Green plants not only take in oxygen and give out carbon dioxide, but also take in carbon dioxide through their leaves and water through their roots to build up their own foodstuffs such as sugars and starch. Oxygen is liberated during this process. Sunlight and chlorophyll, the green colouring matter of plants, must be present. Chlorophyll in the presence of sunlight acts as a catalyst and the whole process, known as photosynthesis, is very complicated indeed. The following simplification is useful but in no way tells us the full story which must be left to a later date.

$$\text{Water} + \text{carbon dioxide} \xrightarrow[\text{chlorophyll catalyst}]{\text{light energy}} \text{carbohydrates (sugars and starch)} + \text{oxygen}$$

It is an interesting point that animals have to take in complex food by mouth whereas plants make their own food from the simple substances carbon dioxide and water.

The carbon cycle

The world's population breathes out millions of tonnes more carbon dioxide than they breathe in, so you might expect the proportion of carbon dioxide in the air to be increasing quickly. This, however, is not the case, the proportion of carbon dioxide in the air is always about 1 part per 3000 parts by volume. Obviously, there must be some ways by which carbon dioxide is taken from the air.

In fact, the generation and removal of carbon dioxide are fairly evenly balanced, the circulation of carbon in this way is called the *carbon cycle.*

The gas is generated by:

(a) Breathing of animals.
(b) Burning of carbonaceous materials such as coal and wood.
(c) Decay of rotting animal and vegetable matter.

The gas is removed by:

(a) Rain water, seas and lakes dissolving it.
(b) Photosynthesis. About 300 000 million tonnes of carbon dioxide are taken in by plants each year.

A look at the process of rusting

When a bicycle, motor car or anything made of iron is left in the open for a considerable time, brown rust often appears on the surface of the metal. Rust is an oxide of iron, and it is interesting to note that while man obtains iron from the naturally occurring ore, haematite, mainly iron oxide, nature effects the reverse process – converting iron to iron oxide.

What causes rusting?

Put two or three clean, shiny nails in each of three test-tubes and treat them as follows:

1st tube: Add water so that the nails are covered.
2nd tube: Add anhydrous calcium chloride to remove water vapour from the air inside the tube.
3rd tube: Add water to cover the nails and then boil the water for a few minutes to expel dissolved air. Then allow vaseline to run down the inside of the hot tube on to the surface of the boiled water. The vaseline sets on cooling and forms a layer on the surface of the water which prevents air entering.

Put a rubber stopper in each of the three tubes and leave them in a test-tube rack for a few weeks inspecting them from time to time (Fig. 25).

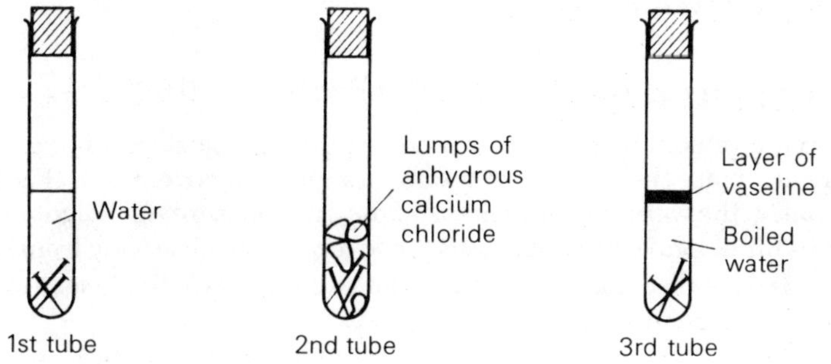

Fig. 25 Rusting of iron nails

As rusting takes place only in test tube 1, we can deduce that both water and air must be present for the iron nails to rust. The way rusting occurs is very complicated, but we now know that water and oxygen are both needed to make iron rust. It was thought at one time that atmospheric carbon dioxide played a part and was necessary for rusting to take place, but this has been shown not to be the case.

How can we prevent rusting?

It has been estimated that in the United Kingdom alone about £50 million is spent each year to help prevent iron from rusting. You will easily see that if we can prevent water and air from reaching the iron then no rusting will take place. As iron is the backbone of industry, used so much in engineering and constructional work, prevention of rusting is very important.

Some ways of preventing rusting are:

1. Painting — used for bridges, vehicles, etc.
2. Oiling and greasing — used for tools and machinery.
3. Coating the surface with a suitable metal.
   (a) Tin. "Tin cans" are made of iron coated with tin.
   (b) Zinc. Roof sheets, buckets and dustbins are dipped in molten zinc and thus coated. The process is called galvanisation.
   (c) Chromium. This is achieved first by electroplating the iron with nickel, then with chromium.
4. Stainless steels consist mainly of iron together with other metals, e.g. chromium (about 12 per cent).
5. Coating the surface with the blue-black oxide of iron – gun barrels, fish hooks and razor blades.

The Forth Road Bridge in Scotland is one of the largest in the world. In its construction 30 000 tonnes of steel have been used and to help prevent rusting a coat of zinc followed by four hard layers of paint have been applied.

What proportion of air is used during rusting?

Attach a muslin bag or cotton wool containing moist iron filings to a wire and place the bag or cotton wool in a tube of air over water (Fig. 26). Equalise the water levels inside and out and measure the height of the air column. Leave the apparatus for several days, observing from time to time. Equalise the water levels again and measure the height of the air column again.

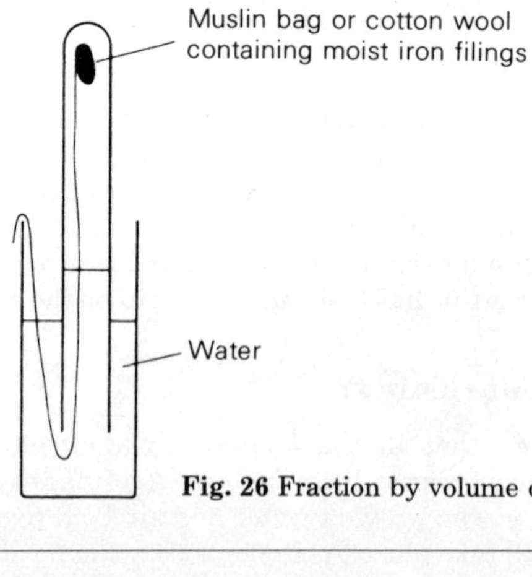

Fig. 26 Fraction by volume of air used in rusting

From the result work out what fraction of air has been used during rusting. What can you conclude from the result?

Burning, rusting and respiration

We have met all three of these processes and they all involve the action of oxygen, although other factors may also play their part.

Burning

In this process, oxygen reacts with the combustible substance giving out heat as well as the products of combustion.

Rusting

This requires not only oxygen but also water along with the iron.

Respiration

This is the process by which a living organism gets energy needed for growth and movement. Sugars, fats and starchy foods all contain carbon and hydrogen and react with oxygen in the body to give carbon dioxide, water and energy, some of the latter being in the form of heat.

THINGS TO REMEMBER

The meaning of the following terms:

- Catalyst
- Basic oxide
- Acidic oxide
- Combustion
- Stalactites and stalagmites and their formation
- The lime water test
- Carbon cycle
- Photosynthesis

QUESTIONS TO TRY

1. Complete the following table:

Substance	Metal or non-metal	Product on burning in oxygen	Type of oxide	Action of water + litmus on oxide
Carbon				
		Sulphur dioxide		
Sodium				
		Calcium oxide		

2. From your knowledge of the reactions of elements in oxygen can you suggest what children's sparklers are made of?

3. Explain how you would carry out an experiment to show that approximately four-fifths of the air by volume does not support combustion.

4. Give the chemical names of:
   (a) quicklime (b) slaked lime (c) marble.

5. Describe and explain what you would see when carbon dioxide is passed through lime water for:
   (a) a short time
   (b) a long time.

6. What is meant by a reversible reaction?

7. Why is the proportion of carbon dioxide in exhaled air greater than in inhaled air?

8. Write a short simple account of:
   (a) photosynthesis (c) uses of oxygen
   (b) the carbon cycle (d) uses of carbon dioxide.

9. Why would you not use dilute sulphuric acid with marble chips to get carbon dioxide?

10. State methods, one in each case, which are normally used to prevent rusting of the following:
    (a) the body work of a motor car
    (b) the bumper bar of a motor car
    (c) a dustbin
    (d) the blades of a lawn mower
    (e) water taps
    (f) food cans
    (g) fishing-hooks.

11. In what ways can burning, rusting and respiration be regarded as:
    (a) similar
    (b) different.

12. Classify the following as physical or chemical changes giving your reasons:
    (a) Smouldering phosphorus in air;
    (b) The "misting up" of a glass of iced water;
    (c) Heating mercury in air;
    (d) Liquefaction of air;
    (e) Production of calcium oxide from limestone;
    (f) Formation of stalagmites and stalactites;
    (g) Extinguishing a fire using carbon dioxide;
    (h) Respiration.

13. Alka Seltzer tablets contain citric acid, aspirin and sodium hydrogen-carbonate. Explain why there is effervescence when a tablet is added to water. What is the gas evolved?

14. When water from chalk or limestone regions is boiled in a kettle a white deposit called "fur" builds up inside the kettle. The addition of acid causes effervescence and removes the "fur". Explain what you think the "fur" might be and what is the gas that is liberated when acid is added to it?

4 What Are Substances Made Of?

It was thought, at one time, that everything was made from FIRE, EARTH, AIR and WATER, these four substances being called elements. Today we know that many more than four elements exist and have a very different picture of what an element is.

The black deposit formed when bread, paper and wood are burned is the same substance whichever substance is burned. This black deposit is carbon and although it will undergo chemical reaction with other substances, the products are always more complex than carbon itself. In fact, carbon cannot be changed into a simpler substance by chemical reaction; it is called an *element*.

Elements are the building bricks of matter. There are over ninety naturally occurring elements and man's invention of the cyclotron and nuclear reactor has made the production of another thirteen possible. We can say that an element is a substance which cannot be split into simpler substances by a chemical change.

Nine elements – gold, silver, copper, mercury, lead, tin, iron, carbon and sulphur – were known to the ancient civilisations and are mentioned in the Bible. It was not until the nineteenth century that the greater number of the now known elements were discovered, as in the Middle Ages the chemists, or alchemists as they were called, devoted most of their energy in trying to convert some of the common metals, like iron, into gold. By the turn of the nineteenth century there were known in all, eighty-four elements.

Abundance of elements

Elements vary considerably in abundance (Fig. 27). In fact nine elements make up about 98 per cent of the whole of the Earth's crust.

The remaining 2 per cent includes well-known elements like copper, zinc and tin. Few elements are found free; most of them are found in combination with one another, e.g. silicon and oxygen as silicon dioxide in sand, sodium and chlorine as sodium chloride in rock salt.

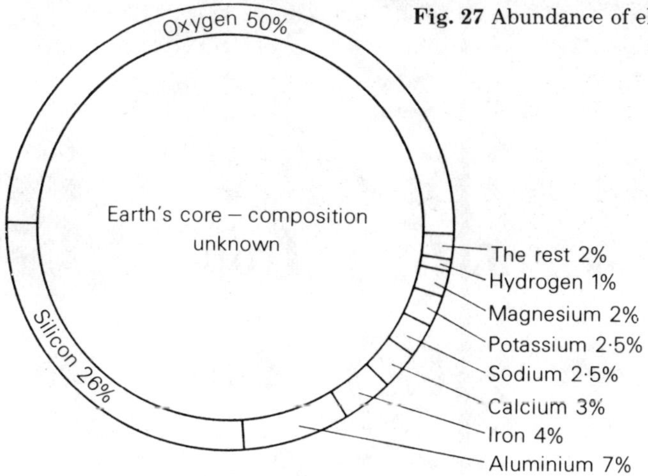

Fig. 27 Abundance of elements

Metals and non-metals

We have seen that by considering the behaviour of the oxides of the elements they can be divided into two groups – those which form basic oxides when burned (or heated) and those which form acidic oxides under the same circumstances, i.e. the metals and the non-metals.

Examples	**Metals**	**Non-metals**
	Copper	Carbon
	Lead	Sulphur
	Sodium	Phosphorus
	Magnesium	
	Mercury	

Metals and non-metals differ in *physical* properties. Metals usually have relatively high densities, are good conductors of heat and electricity and are sonorous (they ring when struck). Non-metals, on the other hand, have relatively low densities, are bad conductors of heat and electricity (indeed they might not even conduct electricity at all), and are not sonorous but often brittle.

Metals and non-metals also show differences in *chemical* properties. As we have seen the former burn in air to form basic oxides while the latter usually form acidic oxides. Metals often liberate hydrogen from dilute acids whereas non-metals never do this.

Plate 3 Dalton's symbols for the elements, 1806–7

Symbols

We have seen on previous pages that chemists like to write down reactions which take place as equations, putting the reacting substances on one side and the products on the other. To make this less laborious and also to give more information about substances, chemists have evolved sign writing, i.e. the names of elements are denoted by letters of the alphabet. This is not an original idea; it was done in the Middle Ages by the alchemists who took the astrologers' signs and gave them to certain metals. For example:

Gold ○ (Sun)   Iron ♂ (Mars)   Copper ♀ (Venus)

When the number of known elements increased there were not enough signs so John Dalton, who worked in the early nineteenth century, introduced his own symbols. For example:

Oxygen ○	Nitrogen ⦶
Carbon ●	Sulphur ⊕

This system was not very satisfactory and probably not easy to remember, so it was eventually replaced by the system of Berzelius, a Swedish chemist, who selected the first letter of the name of each element:

Element	**Symbol**
Carbon	C
Oxygen	O
Sulphur	S

As there are only twenty-six letters in the alphabet and over ninety elements, it was necessary, in some cases, to use two letters for the symbol. For example, carbon, calcium and chlorine all begin with the letter C. It was, therefore, decided that the following symbols should be used:

Element	**Symbol**
Carbon	C
Calcium	Ca
Chlorine	Cl

N.B. The first letter is always a capital and the second letter is always a small one.

Some elements were given symbols which are derived from Latin names because, at that time, Latin, was an international, written language.

Element	**Symbol**	**Latin**
Copper	Cu	*Cuprum*
Iron	Fe	*Ferrum*
Lead	Pb	*Plumbum*

Chemin de fer (road of iron) is the French for railway and our word "plumber" will remind you of the symbol for lead.

A table of the symbols of the more well-known elements

(The words in parentheses are Latin.)

Element	Symbol
Aluminium	Al
Argon	Ar
Arsenic	As
Barium	Ba
Bromine	Br
Calcium	Ca
Carbon	C
Chlorine	Cl
Chromiun	Cr
Cobalt	Co
Copper	Cu
Fluorine	F
Gold	Au (*Aurum*)
Helium	He
Hydrogen	H
Iodine	I
Iron	Fe
Lead	Pb
Magnesium	Mg
Manganese	Mn
Mercury	Hg (*Hydrargyrum*)
Neon	Ne
Nickel	Ni
Nitrogen	N
Oxygen	O
Phosphorus	P
Potassium	K (*Kalium*)
Radium	Ra
Silicon	Si
Silver	Ag (*Argentum*)
Sodium	Na (*Natrium*)
Strontium	Sr
Sulphur	S
Tin	Sn (*Stannum*)
Uranium	U
Zinc	Zn

BEWARE OF THE FOLLOWING:

S	Sulphur and
Na	Sodium
P	Phosphorus and
K	Potassium
I	Iodine and
Fe	Iron
Mg	Magnesium
Mn	Manganese and
Hg	Mercury

Atoms

Although elements are often called the building bricks of matter, the fundamental nature of matter has been a subject of speculation for many hundreds of years. It is all very well to say that ninety-two elements make up matter but what are those elements themselves made up of? The Greeks thought that a substance such as charcoal could be divided into smaller and smaller particles until at last there would be a particle which would resist all further divisions. They named this ultimate particle "atom" which means "not to be cut up". This vague notion persisted until the early nineteenth century when the English chemist, John Dalton, put forward some ideas about elements as a result of experiments that had been carried out at about that time. It had been noted that matter was never "lost". The products (substances left at the end) of a reaction had the same total weight as the reactants (substances at the start) though they were different substances, and that a compound like copper oxide always contained the same percentage of copper no matter how it was prepared.

These facts led Dalton to believe that all matter is made up of atoms which are indivisible particles which cannot be created nor destroyed and that the atoms of a particular element are alike, but different from the atoms of other elements. As we shall see these statements are not entirely true but when they were first put forward they explained the facts that were then known and led to the discovery of newer facts.

Size and structure of atoms

It is true that elements are composed of atoms and that an atom is the smallest particle of an element. Atoms are so small that it is impossible to imagine their size. The following examples should emphasise how extremely small they are.

(a) One hundred million atoms, end to end, in a straight line would measure approximately 1 cm.
(b) If the atoms of a pin head were removed at a rate of 1 atom per second it would take 100 million million years to remove the whole of the pin head.
(c) The number of atoms in a dot of printer's ink is greater than the population of the world.

Since we have been able to "look inside" an atom by using an X-ray microscope, it has been found that atoms themselves are composed of even smaller particles, but these particles are not particles of carbon, sodium, iron or whatever the element is that we are examining, for the same kind of particles are obtained from all atoms. The two most important particles that are found in atoms are called the *electron* and the *proton.*

Electrons and protons

The electron is exceedingly small having practically no weight at all but carrying a negative charge of electricity. No particle has been discovered which carries a smaller charge of electricity so we therefore say that an electron carries unit negative charge or a charge of -1.

The proton has approximately the same mass as an atom of

hydrogen, the lightest atom known, and this is more than 1800 times the mass of an electron. The proton also carries unit electrical charge but in this case the charge is positive. Since there is no electrical charge in the atom as a whole, the negative charges on the electrons must equal the positive charges on the protons and, since each electron and each proton carries one electric charge, the number of electrons in the atom must equal the number of protons, for one negative charge is balanced by one positive charge.

The number of protons, and therefore the number of electrons in an atom of an element is called the *atomic number* of the element.

Neutrons

At Cambridge in 1932 James Chadwick discovered a third type of particle which could be obtained from atoms. This particle, like the proton, has about the same mass as a hydrogen atom but, unlike the proton, it has no charge. He called it a neutron.

Mass number

As the mass of an atom is almost entirely due to its protons and neutrons, the total number of protons and neutrons in an atom of the element is called the *mass number* of the element.

Simple picture of an atom

Investigation of the inside of the atom shows that the protons and neutrons are congregated in the centre of the atom and, since these are the particles with mass, the centre of the atom, or the *nucleus,* as it is called, is the heavy part. The electrons, on the other hand, form clouds of negative electricity encircling the nucleus but some distance away from it.

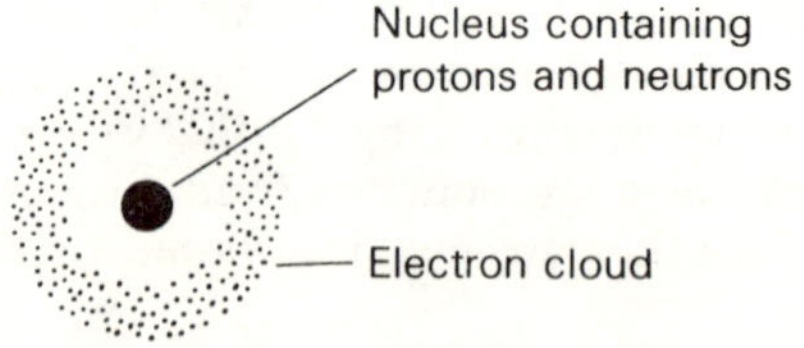

We represent the electrons as a cloud since they are always changing their positions: an electron may be anywhere within its particular cloud at any particular time. We cannot give the exact position of an electron, we can only say where it may be. Our picture of the atom is very different from that of Dalton. We know today that the atom is mainly *empty space* – if the nucleus of an atom were enlarged to the size of a cricket ball, the electrons would be represented by a tiny speck about a kilometre away! All atoms are built on the plan of a positively charged nucleus with negatively charged electrons some distance away.

Summary

	Mass	Electric charge
Proton	About the same as a hydrogen atom	+
Electron	About $\frac{1}{1840}$ th that of a hydrogen atom	−
Neutron	About the same as a hydrogen atom	0

Atomic number of an element is the number of protons in one atom of an element.

Mass number = number of protons + number of neutrons

Compounds

When elements combine together during chemical reaction, *compounds* are formed. Lavoisier, in his classical work on air, heated mercury in air and obtained red mercury oxide formed by the combination of the elements mercury and oxygen.

$$\underset{\text{(element)}}{\text{Mercury}} + \underset{\text{(element)}}{\text{oxygen}} \longrightarrow \underset{\text{(compound)}}{\text{mercury oxide}}$$

Although mercury oxide contains the elements mercury and oxygen, it is not at all like them in appearance or behaviour. Its properties are not even intermediate between those of mercury and oxygen. Mercury, sometimes called "quicksilver" is a "silvery", dense liquid, oxygen is a colourless gas and mercury oxide a red solid.

Common salt, sodium chloride, a white crystalline solid which we sprinkle over our food, is a compound composed of the elements sodium, a light metal which reacts violently with water, and chlorine, a green poisonous gas – a striking example of the difference between a compound and its elements.

Mixtures

The most familiar mixture to us is the air consisting mainly of nitrogen, oxygen, carbon dioxide, water vapour and the noble gases. When the components of air are mixed in the proportions in which they occur in air, air is obtained. No chemical reaction has taken place, there is simply a mixture of elements and compounds and the properties of the mixture are intermediate between those of the components. For example, oxygen is a vigorous supporter of combustion; the other gases do not support combustion. Air is also a supporter of combustion; substances burn in it but not as easily as they burn in oxygen. That is, it behaves as though the oxygen had been diluted which is exactly the case.

Test on iron and sulphur – two elements

Weigh out 7 g of iron filings and 4 g of powdered sulphur and mix them thoroughly. Divide the mixture into two parts, A and B.

Part A

Note the appearance, divide it into three parts and do the following tests:

1. Shake some with water in a test-tube.
2. Add dilute hydrochloric acid to some in a test-tube and warm the mixture. Remove the tube from the flame, put your thumb over the mouth of the tube to allow pressure to build up inside, remove your thumb and immediately put a lighted splint at the mouth of the tube.
3. Hold a magnet near the mixture.

Part B

Heat this mixture in a test-tube and as soon as the mixture becomes hot remove the flame and note what happens in the tube. After cooling, divide the solid into three parts and do the same three tests as in Part A. Finally, carry out the three tests on iron and sulphur *separately*. Write down all your results and see if they agree with what we have already said about mixtures and compounds. Your teacher may let you copy out the table below and ask you to complete it.

	Water	Dilute acid	Magnet
Iron sulphur mixture			
Iron sulphide			
Iron			
Sulphur			

N.B. The black compound formed when iron and sulphur are heated together is iron sulphide. The red glow which persisted after the heat had been removed shows that the reaction is producing heat. We call such reactions *exothermic* reactions.

There is often a heat change when a chemical reaction takes place; either the temperature falls during the reaction or it rises. In the first case when the reaction absorbs heat and the temperature falls if we do not supply some heat by means of a Bunsen burner or other method, we say it is *endothermic*, it takes *in* heat; in the second case, when heat is given out, we say the reaction is *exothermic* (exit = out). This type of reaction, once it is started, often gives out enough energy in the form of heat to keep itself going, e.g. a piece of coal once set alight goes on burning without further need of heat being given to it.

On looking at our results entered in the table above, we now ought to be able to draw up a table showing the main differences between compounds and mixtures.

Main differences between mixtures and compounds

	Mixtures	Compounds
1.	No fixed composition. The components may be in any proportion.	Fixed composition. A given compound always has the elements in the same proportion by weight.
2.	Components may be separated by physical methods.	Component elements can only be isolated by chemical methods.
3.	Properties are intermediate between those of the components.	Properties are very different from those of the components.
4.	Rarely a heat change in their formation.	Often a considerable heat change in their formation.

Molecules

Whilst the smallest particle of an element is called an atom, atoms are not often found free as they join together very readily, and we call the particle so formed a *molecule*.

In the gas oxygen, for example, the atoms of oxygen do not move about on their own but join up in pairs, each pair of atoms being called a molecule of oxygen.

2 atoms of oxygen $\longrightarrow$ 1 molecule of oxygen (usually written O_2).

The same thing happens in the gases hydrogen (H_2) nitrogen (N_2) and chlorine (Cl_2).

In compounds, atoms of different elements join together to form molecules. For instance, in the compound water, a molecule of water is formed by the joining of two atoms of hydrogen to one atom of oxygen. Thus we can represent one molecule of water by the *formula* H_2O. This to the chemist denotes that two atoms of hydrogen are combined with one atom of oxygen.

An *atom* is the smallest possible particle of an element.
A *molecule* is a particle consisting of a cluster of atoms.

Molecules are not all the same size, some contain very many atoms and are therefore quite large compared with an atom. Molecules also differ in shape. While some are spherical, others are long and thin; some are flat like plates, others are twisted. Some consist of hundreds of atoms, like the twisted, thread-like molecules of the proteins of which our body cells are composed. The shape of molecules can determine properties. For example, there is evidence that the shape of a molecule determines to some extent the odour of the substance in question. One theory is that there are seven primary odours: camphor-like, musky, floral, pepperminty, ethereal, pungent and putrid. It is thought that the differently shaped molecules which produce the different odours are distinguished by appropriately shaped receptors in the nose.

What holds atoms together?

Molecules are composed of atoms and it will be obvious that there must be some force which draws the atoms together and holds them there. To answer the question we have to turn again to the structure of the atom and see what happens to one atom in the presence of another.

Basically there are two main types of link, or bond, between atoms, *ionic* bonds and *covalent* bonds. Let us take these in turn.

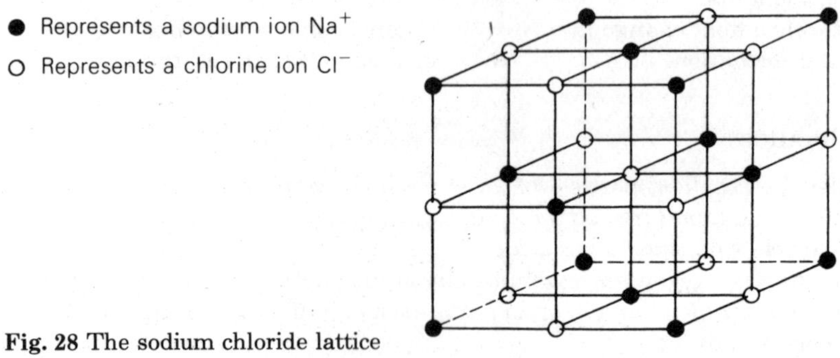

Fig. 28 The sodium chloride lattice

Ionic bonds

An atom is electrically neutral since it carries the same number of positive charges and negative charges but if, in some way, it manages to gain an extra electron, it will now have one more negative charge than it has positive charges and so will become negatively charged. Similarly, if an atom loses an electron, it will have one more proton than electrons and therefore the atom will have an overall positive charge.

This actually happens in some cases and, if we get a negatively charged atom (also called an *ion*) and a positively charged one, then there is a strong electrical attraction between the two ions and they are held together by this force of electrical attraction.

The force of attraction between two ions is called an *ionic bond.* Sodium chloride is an example of a compound in which ionic bonds hold the ions together. The compound is composed of positive sodium ions and negative chlorine ions formed by the transfer of electrons from sodium atoms to chlorine atoms, each sodium atom transferring one electron to a chlorine atom. The ions are arranged in an orderly fashion. Such an arrangement is called a *lattice* (Fig. 28). Many other compounds are also made up of ions but not all of them have the same kind of lattice as sodium chloride. Compounds made up of ions are called *ionic compounds.* They are good conductors of electricity, when molten, and have relatively high melting points and boiling points. Compounds formed from metals and non-metals are usually ionic, the ions present being formed by the transfer of electrons from metal atoms to non-metal atoms.

Covalent bonds

Some elements show no tendency to transfer electrons from one to the other yet they can combine together to form compounds. This is usually the case with two non-metals; take, for example, hydrogen and chlorine. As neither element shows any tendency to lose electrons there is a kind of compromise between the two types of atom. A hydrogen atom and a chlorine atom provide one electron each and this electron pair is then shared by both atoms.

$$\mathbf{H} \overset{\times}{\bullet} \mathrm{Cl} \quad (\overset{\times}{\bullet} \text{ covalent bond})$$

$^{\times}$ Hydrogen electron. $^{\bullet}$ Chlorine electron.

This sharing of electrons is found in many compounds and appears to make a very firm bond between atoms. The bond is called a covalent bond, and compounds containing it are called *covalent compounds.* The bond is generally represented in a formula by a line, thus H—Cl.

Because covalent molecules consist of atoms joined by bonds, it seems sensible to think that because the bonds point in definite directions this would give the molecule a definite shape, and the shapes of a number of molecules which can be observed by the electron microscope have been found to fit in with the direction of the various bonds between the atoms. The direction of the bonds can be found by measuring the angles between them, if there are more than one. In a water molecule, the two hydrogen atoms each share two electrons with the oxygen atom and the bonds between the hydrogen atoms and the oxygen atom are at an angle of 105°.

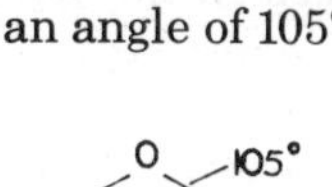

Covalent compounds are usually non-conductors of electricity and have relatively low melting points and boiling points.

Valency and electrons

A. *Ionic compounds*

Here electrons have been transferred from one atom to another. The number of electrons gained (or lost) by each atom is called the valency of the element. In sodium chloride, for example, each Na^+ ion has been formed by the loss of one electron from a sodium atom; each Cl^- ion has been formed by the gain of one electron by a chlorine atom. The valency of both sodium and chlorine is 1. In the case of calcium chloride, each Ca^{2+} ion has been formed by the loss from each calcium atom of two electrons which are taken up by two chlorine atoms to form 2 Cl^- ions. The valency of calcium is 2 while the valency of chlorine is 1.

You will find the following table of ionic valencies useful.

Positive ions		**Negative ions**	
K^+	potassium	CO_3^{2-}	carbonate
Na^+	sodium	NO_3^-	nitrate
Ca^{2+}	calcium	SO_4^{2-}	sulphate
Mg^{2+}	magnesium	Cl^-	chloride
Al^{3+}	aluminium	Br^-	bromide
Zn^{2+}	zinc	I^-	iodide
Fe^{2+}	iron (II)	OH^-	hydroxide
Fe^{3+}	iron (III)	HCO_3^-	hydrogencarbonate
Pb^{2+}	lead		
Cu^{2+}	copper		
Ag^+	silver		
NH_4^+	ammonium		

B. *Covalent compounds*

In these cases the valency of an element is the number of electrons which an atom of the element has provided for sharing purposes with other atoms. In water, for example, each oxygen atom provides two electrons to form covalent bonds with the two hydrogen atoms. Each hydrogen atom provides one electron for sharing. The valency of oxygen is two; the valency of hydrogen is one.

Changes of state

It seems almost unbelievable that a solid like ice should consist of exactly the same substances as the liquid water and the gas water vapour; but ice, water and water vapour all contain "clusters" of two hydrogen atoms and one oxygen atom, i.e. H_2O "clusters" or molecules. It might be a good idea to have a closer look at the states in which matter can exist, i.e. SOLID, LIQUID and GAS.

Solids

Solids are usually described as hard, at least compared with liquids, and as having a definite shape. A powder does not appear to have a definite shape; we can make it into a pointed heap or a flat heap, but in a heap of powder we are looking at quantities of grains each of which has a definite shape. If one examines a little common salt under a microscope it will be seen to consist of tiny cubes, some of them damaged, may be, but there will be some perfect ones. Epsom salts consist of tiny needles and alum consists of octahedrons. All these substances have a fixed shape and we call them crystals. Thus we can say that a solid is a crystalline substance. It is not always easy to see the crystals, even through a microscope, and at one time it was thought that a powder such as chalk was amorphous, a word which means without shape. However, we now know that chalk is a crystalline substance and even metals are made of exceedingly tiny crystals packed very close together.

Now, we know that a crystal of common salt is composed of millions of sodium ions and chloride ions held together by electrical attraction, and since chemists have been able to investigate a very tiny crystal by means of X-rays, it has been discovered that the sodium and chloride ions are always arranged in a very regular pattern in relation to one another. In fact each sodium ion is surrounded by six chloride ions, and each chloride ion is surrounded by six sodium ions except at the edge of the crystal. If you try to make an arrangement like this with tiny balls of two colours of plasticine to represent the sodium and chloride ions and matchsticks to hold them together, you will find that you can build a cube and this will be a representation of a sodium chloride crystal. A crystal of another salt, say silver nitrate, would still be built up to a pattern, but in this case the pattern would be different from the sodium chloride one. A pattern of this kind is called a crystal lattice (see Fig. 28) to which we have already referred. Silver nitrate crystals have a different lattice which produces the differently shaped crystal of the salt.

A crystal composed of ions is called an *ionic crystal*. True molecules as well as ions are able to form crystals and these crystals are known as *molecular crystals*. Naphthalene is an example of a molecular crystal.

Giant molecules

Some molecules are very large indeed compared with simple molecules like hydrogen (H_2), oxygen (O_2) and hydrogen chloride (HCl). Each of these very large molecules consists of hundreds of atoms. They are so large in comparison with the simple molecules that they are called *giant molecules* or *macromolecules*.

Diamond, a form of carbon, is an example of a macromolecule. The carbon atoms are arranged tetrahedrally which means that each

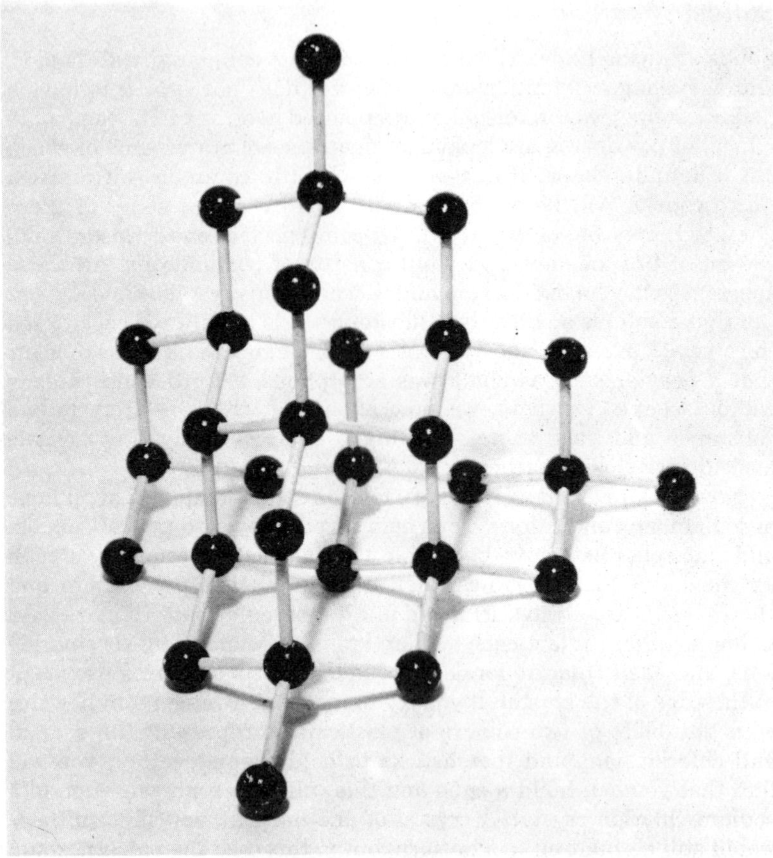

Plate 4 Model of diamond structure

carbon atom is at the centre of a tetrahedron and attached to four other carbon atoms which are at the four corners of the tetrahedron (see Plate 4). The structure of diamond is very compact and accounts for the hardness of this particular form of carbon.

Graphite, another form of carbon, is another example of a macromolecule, but in this form the atoms are arranged differently from diamond (see Plate 5). The carbon atoms are in the form of layers of hexagons. The structure is much less compact than that of diamond and accounts for the relative softness of graphite. So called "lead pencils" are made of graphite and you know how easy it is for these pencils to be sharpened. Other examples of macromolecules include nylon (hexamethylene-tetramine) (see Plate 6) and DNA (deoxyribonucleic acid) (see Plate 7), the basic genetic material. Your teacher may show you models to illustrate some macromolecules.

Plate 5 Model of graphite structure

Plate 6 Crystal structure models of sodium chloride, diamond, graphite, garnet, hexamethylene-tetramine and hexamethyl benzene.

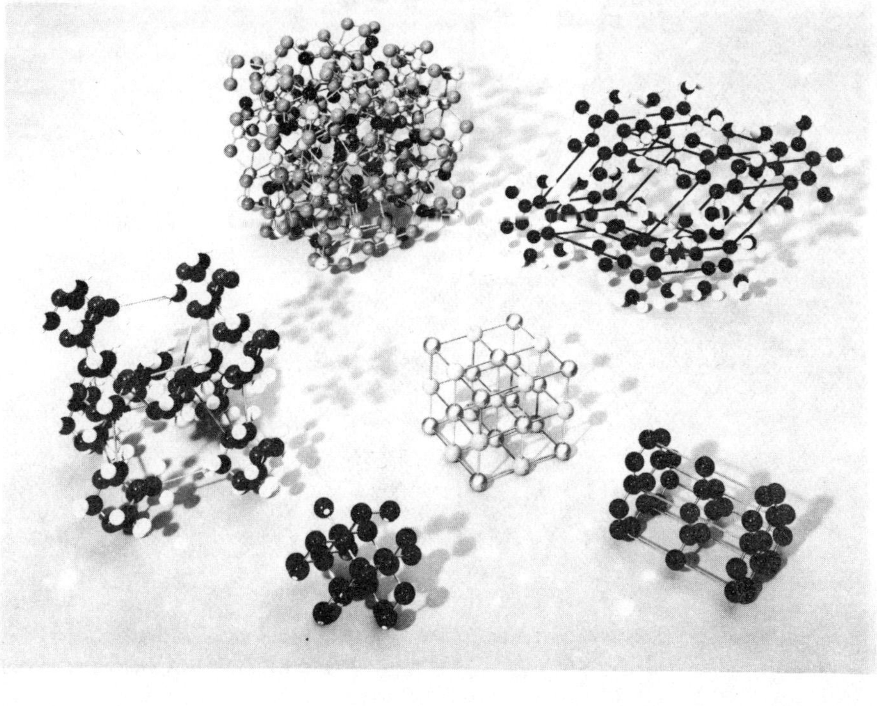

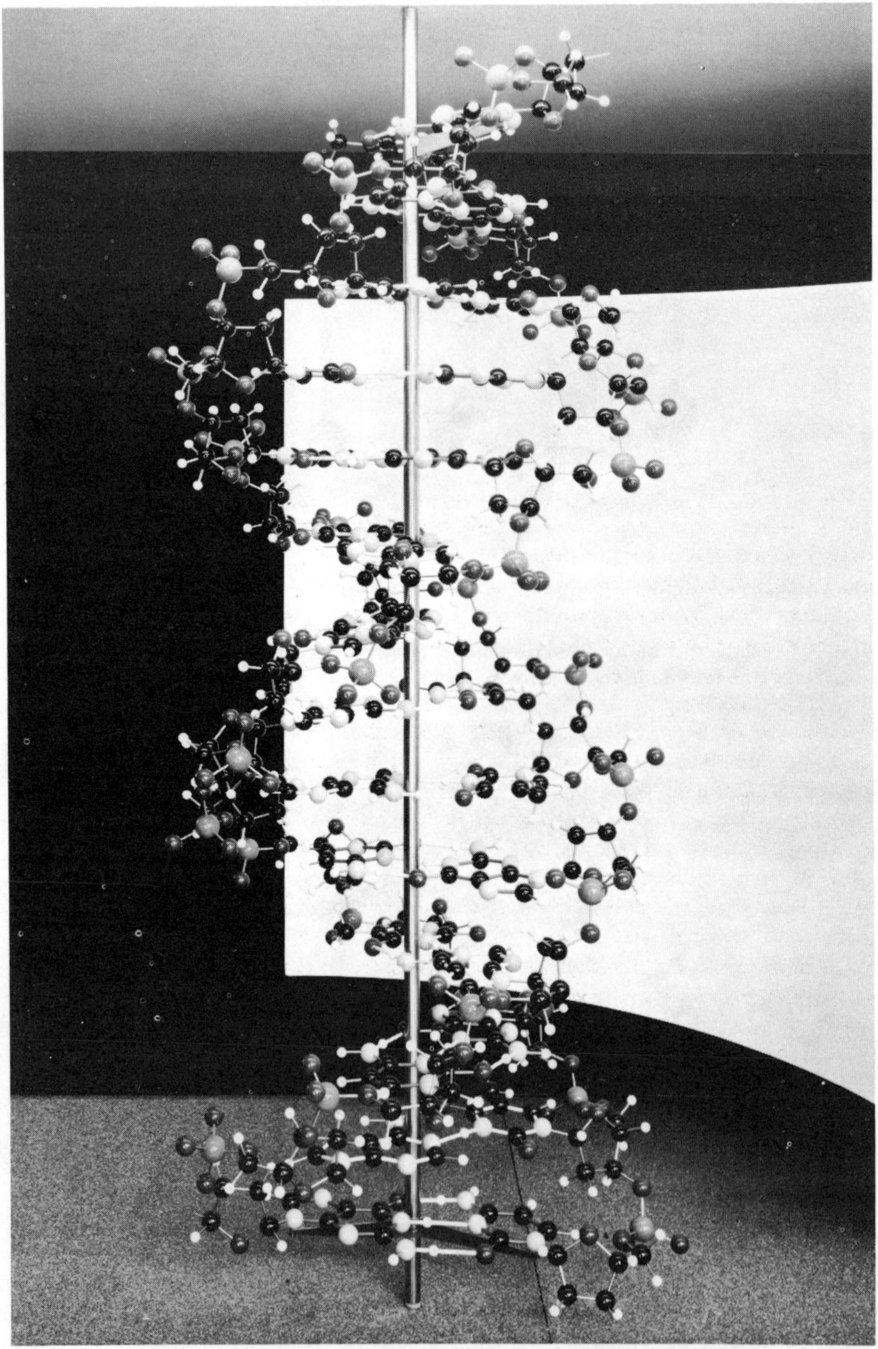

Plate 7 The DNA molecular model

The forces holding the ions or molecules together are quite strong when these particles are near to each other. This is particularly so in the ionic crystal where the force is electrical attraction, and so we find that ionic crystals are generally much harder than molecular ones.

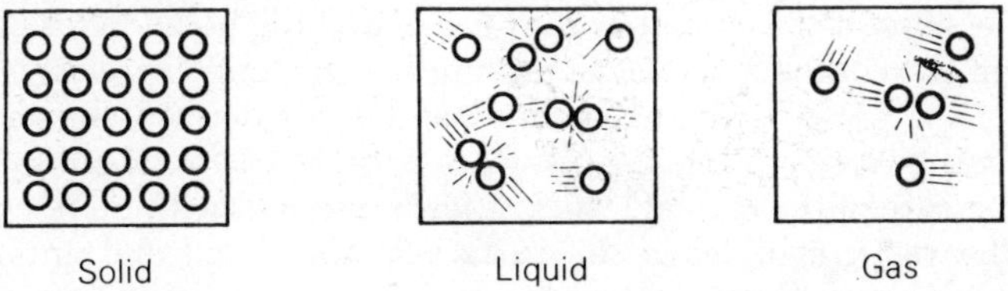

Particles vibrate to and fro. No *free* movement

Particles slide freely within the liquid. Frequent collisions

Particles far apart

Fig. 29 States of matter

We do not think of the particles that make up a solid as moving since a solid does not move on its own accord, but they do actually move very slightly to and fro like a pendulum, though not enough to change their place in the lattice (Fig. 29). We say they are vibrating slightly. Now heat is a form of energy and when a solid is heated the particles of a solid are given more energy which makes them vibrate more and more until at last they are able to overcome the force of attraction between themselves and so break away from each other. They begin to move around, but as there is still a certain amount of attraction between them the movement is not sufficient to make them move very far from each other. When all the molecules reach this point the substance melts and becomes a liquid. We said that the forces holding the ions together in an ionic crystal were very much stronger than those holding the molecules together in a molecular crystal, therefore the ionic crystal has to have much more energy in order to overcome these forces; that is the ionic crystal melts at a higher temperature than does the molecular crystal and we say it has the higher melting point. When the source of heat is taken away from the crystals, the molecules gradually lose their energy, the forces of attraction get stronger and as the molecules can no longer travel the greater distances they are drawn into the crystal lattice again and the liquid *solidifies* or *freezes*.

If we look at a crystal lattice, for example the common salt lattice, it can be seen that while most of the ions are each surrounded by six ions of a different polarity or charge, those at the surface of the lattice are not so well guarded. In the case of molecular compounds where the force between the molecules is not so great, these surface molecules sometimes have enough energy to break away. In this case they form a vapour and we say that the solid is *volatile*. Sometimes we get a few molecules in our nostrils and we say we can smell the solid. This is the case with naphthalene.

Liquids

We have seen that as we heat a molecular solid its molecules gain energy until they are able to break out of the lattice and the solid melts. The molecules have not yet enough energy to move very far apart, but they are free to move. When they do so, they often collide with the sides of the container they are in and the liquid takes up the shape of the container. If not in a container the liquid spreads out on the surface – that is the substance has lost its defined shape. In a container, a liquid still has one surface open to the atmosphere and, as in the case of the crystal lattice, a few molecules with more energy than the rest might leave the surface of the liquid and enter the atmosphere as a vapour. Since the molecules in a liquid as a whole have more energy than those in the crystal lattice and since the force of attraction between the molecules is already less strong, it is probable that more of them will escape and thus more liquids than solids give off smell, e.g. perfumes, methylated spirit, eucalyptus oil, vinegar, brandy, etc.

If we heat a liquid we are providing even more energy and the molecules begin to move about with much greater vigour. Thus, many more molecules reach the surface of the liquid and, having more energy, they overcome the force of attraction and leave the liquid as vapour. When eventually all the molecules begin to behave in this manner the liquid *boils* and the temperature at which this takes place is the *boiling point* of the liquid. The opposite occurs if a hot vapour is cooled. The molecules lose their energy, begin to move shorter distances and to move more slowly, the attraction between them grows stronger until a point is reached where they can no longer remain free. They move together forming small groups and drops of liquid are formed. *Condensation* has taken place.

Gases

As we have seen above, when a solid or liquid is heated, there comes a time when the molecules have got enough energy to break away from the intermolecular attraction forces and become free. Now they can move in all directions and will continue to do so until stopped by a collision, either with another vapour molecule or with the surface of a liquid or solid. When this happens, the molecules rebound, like a rubber ball, and go off in another direction until the next collision takes place. This means that the gases have no fixed volume, they fill all the space that is available and there will be a continual bombardment of the walls of the container by the molecules of gas (see Fig. 29). Since many millions of molecules are continually carrying out this bombardment a considerable force is brought to bear on the walls, and the force per unit area is known as the pressure exerted by the gas.

Molecules in motion – an experiment to try

The following experiment is a very simple one. The apparatus and substances needed are inexpensive and the practical work can be carried out at home without any difficulty.

Fill a gas jar or milk bottle to the brim with water. Then pour a little ink into another gas jar or milk bottle so that the bottom of the vessel is just covered and pour water into the vessel so that it is filled to the brim.

Slide a piece of cardboard or postcard over the mouth of the vessel containing water and, taking care to hold the cardboard in position so that no water runs out, invert the vessel and place it mouth to mouth with the other vessel containing the ink and water. Now that the two mouths of the vessels are together, carefully pull out the cardboard so that the liquids in the two vessels can meet. Let the apparatus stand for several hours and from time to time have a look at the results.

At the same time as you set up the apparatus described above, you can also carry out the same experiment but with the vessel containing the ink and water resting on top of the vessel containing water only. Again let the apparatus stand for several hours and observe what happens from time to time.

From what you know about liquids can you explain what has happened during the above experiments?

Another experiment to try

Half fill a glass vessel such as a beaker or jam jar with water and add a crystal of a coloured salt such as copper sulphate or potassium permanganate but do not stir. Leave the apparatus for several hours and note what happens. Can you explain the results?

Diffusion

At some time or other you will have noticed odours, pleasant and unpleasant, on entering a chemistry laboratory. The source of the smell may have been just a small reaction flask but the odour soon "spreads" itself out and mixes with the air so that it can be detected anywhere in the laboratory. The beautiful odour of flowers often tells us that we are near a garden well before we actually see it. Why is this so? We know that gases are made up of molecules which can move very quickly indeed, and odours travel fast because the molecules which cause the odours can travel very quickly. The molecules causing the odour spread themselves out and mix with the molecules of air which, of course, are also moving in random fashion.

The intermingling of gases as a result of the movement of their molecules is called *diffusion*.

When a few drops of the dense brown liquid, bromine, are placed at the *bottom* of a gas jar by means of a pipette, and a cover is placed over the jar, the brown fumes of bromine soon spread out in the jar, mixing with the molecules of air and rising up in the process *even though bromine is much denser than air.*

Diffusion is not confined to gases. Liquids will also diffuse and you will be able to explain the result of the experiment with ink and water by the intermingling of the molecules of ink and water. In the experiment with a coloured crystal and water, the crystal is bombarded by water molecules and a solution is slowly formed around the crystal. As time passes, the coloured solution spreads itself by diffusion until the whole solution is evenly coloured. Of course, colourless crystals will also undergo solution in the same way but the diffusion cannot be as easily observed.

To find out if gases diffuse through a porous pot at the same rate

Fig. 30 Diffusion, using air and carbon dioxide

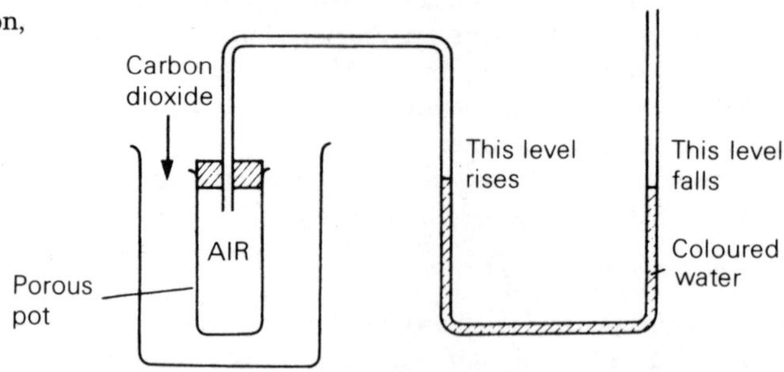

Using the apparatus shown in Fig. 30, lead carbon dioxide gas into the glass beaker and note what happens. Take the beaker away and again note the effect.

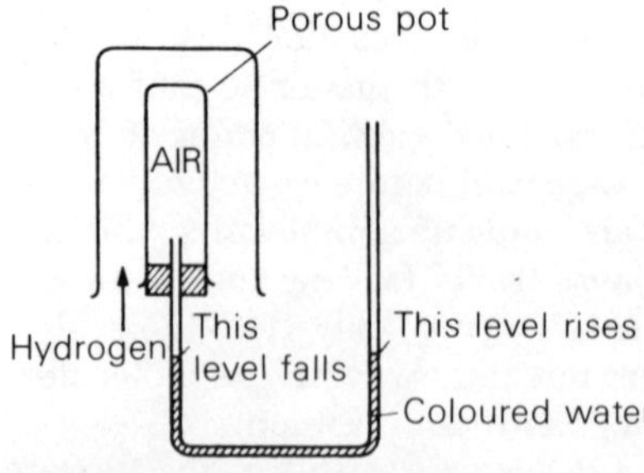

Fig. 31 Diffusion, using air and hydrogen

Using the apparatus shown in Fig. 31, lead hydrogen gas from a cylinder into the beaker and note the effect. Then remove the beaker and see what happens.

You will notice that the levels of coloured water move in a way which shows that the pressure inside the porous pot decreases when the pot is surrounded by carbon dioxide. However, when the pot is surrounded by hydrogen gas the water levels move in such a way as to indicate that the pressure in the pot is increased. We can explain this as follows: carbon dioxide gas is more dense than air and while molecules of both these substances pass through the porous pot, the carbon dioxide molecules being more dense pass through more slowly. A drop in pressure inside the porous pot results and water rises up the left tube.

On the other hand, hydrogen gas is less dense than air and its molecules pass through the porous pot more quickly. An increase in pressure inside the porous pot results and the water level falls down the left tube and rises up the right tube.

Diffusion has been used to separate different kinds of uranium atoms for atomic energy purposes (see Chapter 8).

Structure of matter

PROTONS NEUTRONS ELECTRONS

ATOMS – these rarely exist uncombined. A combination of atoms is called a molecule.

MOLECULE

In molecules of ELEMENTS the atoms are all alike, e.g.
- Oxygen O_2
- Nitrogen N_2
- Phosphorus P_4

In molecules of COMPOUNDS the atoms are different, e.g.
- Water H_2O
- Carbon dioxide CO_2
- Tetrachloromethane CCl_4

THINGS TO REMEMBER

Can you give the following definitions? If not look them up and learn them.

- An element is ...
- An atom is ...
- Mass number is ...
- Atomic number is ...
- A compound is ...
- A mixture is ...
- A molecule is ...
- The three states of matter are ...

QUESTIONS TO TRY

1. Write the symbols for the following elements: magnesium, manganese, mercury, sodium, zinc, copper and lead.
2. What elements are represented by: K, P, Ag, Fe, Cr, Ca?
3. Write down the formulae of:
   (a) sodium carbonate
   (b) potassium nitrate
   (c) lead chloride
   (d) aluminium sulphate
   (e) calcium hydroxide
   (f) copper nitrate
   (g) ammonium sulphate
   (h) magnesium hydrogencarbonate
   (i) zinc sulphate
   (j) iron (III) nitrate.
4. An element X forms compounds of formulae XO, XCl_4, XO_2, XSO_4, $X(NO_3)_2$. What are the two valencies of the element X?
5. An element Z has the valencies 3 and 5. Write down the formulae of the two chlorides and the two oxides.
6. Copy the following statements in your book and fill in the gaps:
   (a) When metals combine with non-metals, the metal atoms usually ... electrons and the non-metals usually ... electrons. The compounds formed are called ... compounds and usually have ... melting points and boiling points and, when molten, the compounds are ... conductors of electricity.
   (b) When non-metals combine with one another, the atoms are held together by a . . . bond, which consists of a . . . of electrons between the atoms. The compounds formed usually have relatively ... melting points and boiling points and are ... conductors of electricity.
7. Give three reasons why you think:
   (a) air is a mixture (b) sodium chloride is a compound.
8. Explain the following:
   (a) A hot meal can be detected by its odour more quickly than a cold meal.
   (b) A moth ball left in some clothes for several months "disappears".
   (c) Water evaporates more quickly from a saucepan than from a milk bottle.

5 The Action of Electricity on Substances

Electricity contributes a great deal to our everyday life. Obvious uses include heating and lighting, but its application to chemistry is not so well known. In this chapter we shall have a look at the changes which take place when an electric current is passed through substances. This branch of chemistry is called *electrochemistry.*

Conductors and non-conductors

The first question to ask ourselves is, "Do all substances allow electricity to be passed through them?" We can immediately think of one example of a substance which will – in the ordinary electric light bulb the metal filament allows the electricity to pass through it and becomes white hot in the process. The extremely hot glow accounts for the light which fills the room. When the electric current is switched off, the metal filament cools down and no more light is given out. The metal filament has not changed at all. Such substances which allow an electric current to pass through them without undergoing any chemical change are called *conductors.*

It is necessary for safety purposes to cover wires which carry an electric current with some suitable material which will not allow an electric current to pass. You will have noticed the plastic sheathing round electric wires or the plastic material used for plugs and sockets. Substances like these plastics which will not allow an electric current to pass through them are called *non-conductors or insulators.*

Electrolytes

Let us now have a look at the substance lead bromide, which at room temperature is a crystalline solid. Using the apparatus (Fig. 32) in a

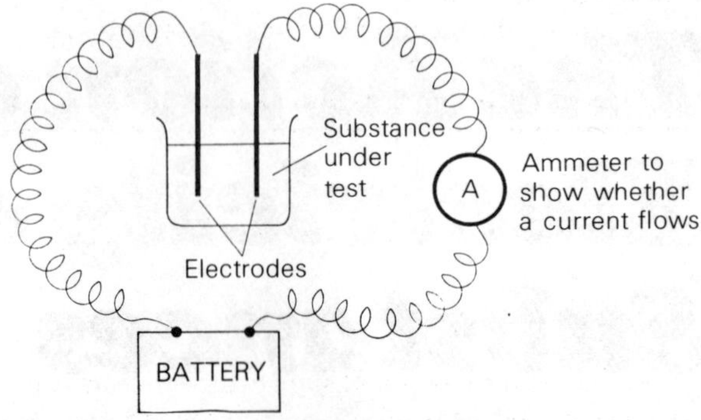

Fig. 32 Action of an electric current on a substance

fume cupboard we can try to pass an electric current through the solid but without success. Solid lead bromide is, therefore, a non-conductor. When lead bromide is heated it melts and the reading on the ammeter shows that it will now allow an electric current to pass through it. Molten lead bromide is, therefore, a conductor, but this is not by any means the whole story. A bead of metal lead and brown bromine vapours can be seen as a result of the passage of the current.

$$\underset{\text{(molten)}}{\text{Lead bromide}} \xrightarrow[\text{(30 V d.c. carbon electrodes)}]{\text{electricity passes}} \text{lead and bromine}$$

A substance which will not only allow an electric current to pass but will also be decomposed by it is called an *electrolyte*.

The process is called *electrolysis*.

The terminals by which the current passes in and out of the electrolyte are called *electrodes*.

The +ve electrode is called the *anode*.

The −ve electrode is called the *cathode*.

Experiments with substances to find out whether or not they are electrolytes

The apparatus in Fig. 32 can be used and the following substances are suggested:

Distilled water	Silver nitrate solution
Dilute sodium hydroxide solution	Tap water
Copper sulphate solution	Sugar solution
Zinc sulphate solution	Dilute sulphuric acid

It is important to wash the electrodes with distilled water after each test solution has been used.

Make out a table as shown and fill in your observations:

Substance	Electrolyte or not	Action at cathode	Action at anode

You will find that acids, bases and salts, when molten or in aqueous solution, are electrolytes. Most other substances are non-conductors.

The gas at the cathode

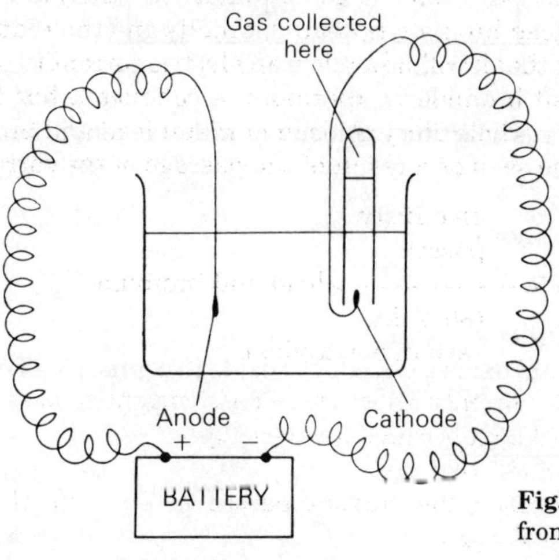

Fig. 33 Collection of gas from the cathode

At the cathode there is always either a metal deposit *or* bubbles of gas liberated. We can collect this gas (Fig. 33) and find that when a lighted splint is applied to it a slight explosion occurs. This gas is called *hydrogen* and one of the tests for it is to put a lighted splint into it and the gas will react explosively with the oxygen of the air to give water. There is often a resounding "pop".

$$\text{Hydrogen} + \text{oxygen} \longrightarrow \text{water}$$

It is worth remembering that during electrolysis:

1. Metals and hydrogen are liberated at the cathode.
2. Non-metals are liberated at the anode.

What sort of gas is hydrogen?

Although the gas can be obtained at the cathode during the electrolysis of a number of substances, a more convenient method of collection is shown in Fig. 34. Zinc will displace hydrogen from dilute sulphuric acid and the gas can be collected over water.

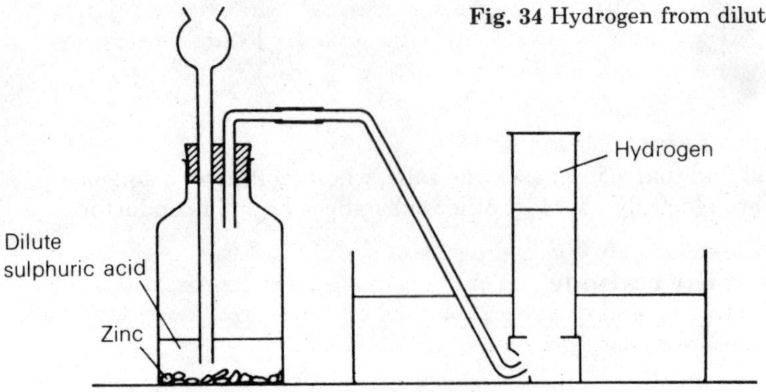

Fig. 34 Hydrogen from dilute acids

In addition to the gas reacting explosively with oxygen, it has the following properties:

1. The lightest gas known.
2. Does not support combustion.
3. Neutral.
4. Only slightly soluble in water.
5. It will reduce metal oxides, e.g. copper oxide to copper; lead oxide to lead.

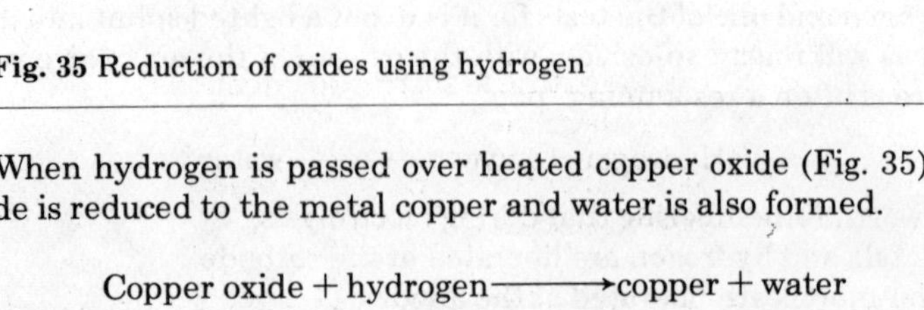

Fig. 35 Reduction of oxides using hydrogen

When hydrogen is passed over heated copper oxide (Fig. 35), the oxide is reduced to the metal copper and water is also formed.

$$\text{Copper oxide} + \text{hydrogen} \longrightarrow \text{copper} + \text{water}$$

Soap bubble explosions

When many aqueous solutions of electrolytes are dilute enough, in addition to the liberation of hydrogen at the cathode there is often oxygen liberated at the anode. For example, when dilute acid is electrolysed the hydrogen liberated at the cathode and the oxygen liberated at the anode can be mixed in the same outlet tube and led into a soap solution. Soap bubbles containing hydrogen and oxygen can be formed in this way and, after catching them on a wooden ruler, can be put into a flame. A loud bang results.

How does electrolysis work?

The following is a simplified picture of what occurs. In the last chapter we said that ions are formed when electrons are added to or taken from atoms. Ions are charged particles and if a substance is made up of ions then, if it is in solution or if it is molten, electrolysis is possible. The electrodes simply attract the ions and once the ions reach the electrodes their charges are neutralised (Fig. 36).

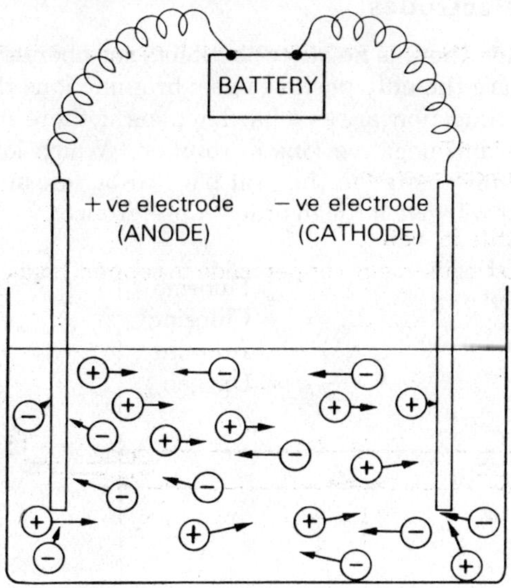

Fig. 36 Ions in an electrolyte

CATHODE **attracts** POSITIVE IONS (cations)
(−ve electrode)

ANODE **attracts** NEGATIVE IONS (anions)
(+ve electrode)

Let us consider lead bromide again. The solid does not conduct an electric current because the ions of lead and bromine are in relatively fixed positions in the crystal lattice. In the molten state, however, or in aqueous solution, the lead and bromine ions are free and can be attracted to the respective electrodes.

+ve lead ions are attracted to the −ve cathode
−ve bromine ions are attracted to the +ve anode

At the cathode:

$$Pb^{2+} + \underset{\text{(from cathode)}}{2\,e^-} \longrightarrow Pb \quad \text{lead deposited.}$$

At the anode:

$$2\,Br^- \longrightarrow Br_2 + 2\,e^- \quad \text{bromine and electrons liberated.}$$

If the ions are not free to move in a substance or if there are no ions at all, as in covalent substances, an electric current cannot be passed.

Competition at the electrodes

In the case of lead bromide there is no other possibility for liberation of other ions. Lead ions are the only positive ones; bromine ions the only negative ones. The question arises what happens if there are several kinds of positive and negative ions in solution? Which ions are liberated first? The full answer to this will have to be left until later but the following list will give a rough order of preference:

Cathode	Anode
Potassium	Fluorine
Sodium	Chlorine
Magnesium	Bromine
Aluminium	Oxygen
Zinc	
Iron	
Tin	
Lead	
Hydrogen	
Copper	
Silver	
↓ *Increasing ease of liberation at the cathode*	↓ *Increasing ease of liberation at the anode*

Electroplating

Coating a key with copper

With the aid of sandpaper clean the surface of a key or some other metal object you want to plate. Attach it to a wire and connect it to the negative electrode of a dry cell. You can use a strip of copper, or a carbon rod from old dry cells and attach it to the positive terminal of the dry cell. Dip the key and the strip of copper (or carbon rod) into a beaker of copper sulphate solution and leave them for several minutes. Then rinse the plated object in water (Fig. 37).

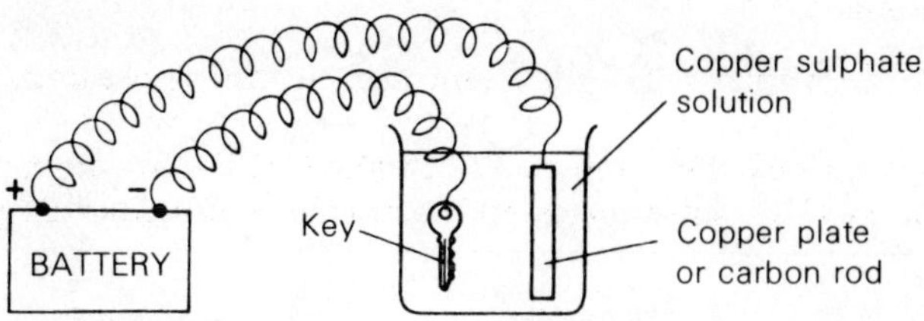

Fig. 37 Electroplating a key

In industry, of course, electroplating is carried out on a very large scale. Silver, nickel, gold and other metals can be electroplated on to objects, which are made the negative electrodes in the electrolysis. In addition to electroplating, electrolysis is also used to make chemicals such as sodium hydroxide and household bleach and to extract metals from their ores, e.g. aluminium, magnesium and sodium, and to purify copper.

The story of aluminium extraction

Aluminium is the most abundant metal in the Earth's crust and, on this score, we would expect it to be cheap. However, much of it is present in clays and silicate rocks which have so far proved two unprofitable sources of the metal.

The first commercial production of the metal, in 1854, was so expensive that the cost of the metal was as high as £7 per pound. One of the chief reasons for this was that the oxide of aluminium, unlike the oxide of iron, could not be reduced to the metal by carbon. However, in 1886, Paul Heroult in France and Charles Martin Hall, a 22-year-old student in America, discovered independently a method of extraction using electricity. The result was a staggering drop in the price of aluminium to 2s 6d (12½ p) per pound. The metal which was once used in place of silver for cutlery on special royal occasions was no longer precious.

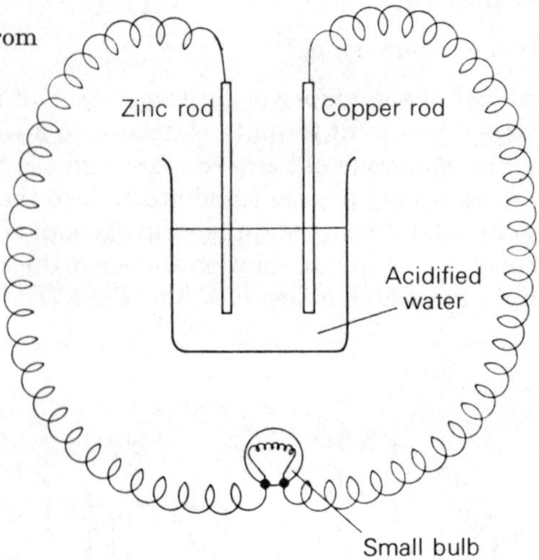

Fig. 38 Electricity from chemicals

Can we make electricity from chemicals?

Put one zinc plate and one copper plate in a beaker of acidified water (Fig. 38). Does the lamp light? Make another "cell" just like the first and join the two cells "in series", i.e. join the zinc plate of one to the copper plate of the other. Does this affect the lamp? Try a third cell and join the three in series and see the effect on the lamp.

THINGS TO REMEMBER

- A conductor is a substance which will allow an electric current to pass through it.
- A non-conductor will not allow an electric current to pass through it.
- An electrolyte is a compound which, in solution or in the molten state, will not only allow an electric current to pass through it but will also be decomposed by it.
- Electrolysis is the process of passing an electric current through a substance to decompose it.
- An electrode is the terminal by which an electric current enters or leaves an electrolyte in electrolysis.
- A cathode is the negative electrode.
- An anode is the positive electrode.
- Metals and hydrogen are usually liberated at the cathode.
- Non-metals (except hydrogen) are usually liberated at the anode.
- An ion is a charged particle. A cation is positively charged. An anion is negatively charged.

QUESTIONS TO TRY

1. Common salt (sodium chloride) has the arrangement of ions in its crystal lattice as shown in Fig. 28.
   (a) Will the solid allow an electric current to pass through it?
   (b) Will the molten salt?
   (c) Will the aqueous solution?
   Give reasons. Can you suggest what, if any, the products will be?

2. Why is it safe to hold an electric plug made of plastic even though a "live wire" may be attached to it?

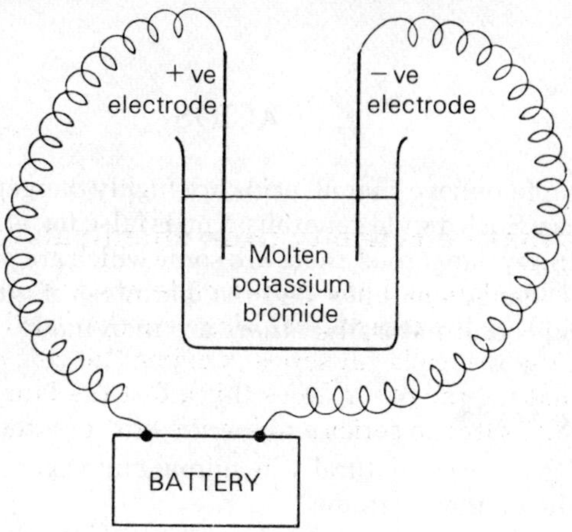

Fig. 39

3. Study the diagram (Fig. 39) and answer the questions:
   (a) What are the ions present in the beaker?
   (b) The positive electrode is called ...
   (c) The negative electrode is called ...
   (d) Which ions go to the positive electrode?
   (e) Which ions go to the negative electrode?
   (f) What is the process called?
   (g) What is the molten potassium bromide called?
   (h) What is the name given to the anode and cathode?
   (i) Apart from switching off the current how could you stop the current passing through the potassium bromide?
   (j) What would happen if the two electrodes were switched to different poles of the battery?

6 Acids, Bases and Salts

ACIDS

Some people believe that all acids are highly dangerous and corrosive substances. Such a wide generalisation is false for, although some acids are extremely dangerous, there are some which are perfectly harmless. Even the dangerous ones become harmless if sufficiently diluted. Some acids occur naturally; others are man-made.

Organic acids

Some of these occur naturally in animal and vegetable matter and all contain the element carbon.

Examples:

Acid	Source
Citric acid	Lemons and oranges
Malic acid	Unripe apples
Tartaric acid	Pears and grapes
Oxalic acid	Rhubarb leaves

Acids account for the sharp taste of certain fruits such as lemons and unripe apples. The word "acid" is from the latin *acidus* meaning sour. The sting of nettles and ants owes its sharpness to formic acid, another name from latin, *formica* meaning ant. Formic acid was first obtained in the seventeenth century by Magraeff when he distilled red ants. Much improved methods are used today. The sharpness of vinegar is caused by acetic acid. Oxalic acid is a poisonous acid and, therefore, rhubarb leaves should not be eaten.

Mineral acids

These acids are manufactured from minerals and include hydrochloric acid, sulphuric acid (used in car batteries) and nitric acid.

Indicators

When washing dishes with bilberry juice on them you will no doubt have noticed that the red bilberry colour turns blue when soap powder is present. If you add vinegar to a little of the blue liquid formed, the colour change is reversed, i.e. blue to red.

Try using firstly beetroot juice and then red cabbage juice on washing soda. See what effect vinegar has on the products.

Test on (a) *bilberry juice,* (b) *beetroot juice,* (c) *red cabbage juice*

Place eight test-tubes in a rack and pour in the following solutions:

1. Lemon juice
2. Vinegar
3. Hydrochloric acid
4. Sulphuric acid
5. Nitric acid
6. Household ammonia
7. Bicarbonate of soda
8. Washing soda

Add a few drops of bilberry juice to each of them and record any colour change. The experiment can be repeated with fresh supplies of the eight solutions but using beetroot juice and/or red cabbage juice instead of bilberry juice.

Some of the eight solutions will turn the colouring matter used red; some will turn it blue. What you have done is a simple test for two important classes of chemical compounds called ACIDS and ALKALIS. Acids turn the colouring matter red, alkalis turn it blue. The colouring matter is called an INDICATOR.

Universal indicator

The colouring matter of fruits and vegetables, although useful as a rough indicator, will not tell us the "amount" of acidity or alkalinity, i.e. whether the substance under test is weakly or strongly acid or alkaline. Universal indicator is very useful because it gives us some idea of the strength of acids and alkalis.

Test

Put some white vinegar into a test-tube to a depth of about 2 cm. and add a few drops of universal indicator. Notice the colour change. Add a few drops of household ammonia, then some more and so on, recording the changes in colour of the indicator, until no further colour change occurs.

Neutralisation

When a farmer wants to cure the "sourness" in the soil he adds lime. We say he is *neutralising* the acid of the soil with lime, which is called a *base*.

N.B. BASES neutralise ACIDS
ACIDS neutralise BASES

When a base is soluble in water it is called an alkali.

What really happens when a base is added a little at a time to an acid is that it gradually makes the acid weaker until there is no acid left. In the experiment with universal indicator the colour is red at first but changes from one colour to another as the base is added. The indicator tells us how strong an acid or a base is.

The pH scale

The range of acidity or alkalinity is measured by what is known as the pH scale. A number, called the pH number, is given to a solution in which the universal indicator has a particular colour, for example:

Colour of indicator	**pH number**
Red	4
Orange	5
Yellow	6
Green	7
Blue	8
Indigo	9
Violet	10

For example, if we added a few drops of universal indicator to a solution and the indicator became green we would say that the solution has a pH number of 7.

The lower the number the more acidic the solution.

The higher the number the more alkaline the solution.

Other common laboratory indicators

Litmus (from lichens), methyl orange and phenolphthalein are all used as indicators in the laboratory. Their colour changes in solutions of different pH number are:

Litmus	pH	below 7	– **red**
		above 7	– **blue**
Methyl orange	pH	below 4	– **red**
		above 4	– **yellow**
Phenolphthalein	pH	below 8	– **colourless**
		above 10	– **red**

What have all acids in common?

So far we have established that acids can change the colour of indicators, the colour change depending on the strength of the acid. There is, however, one thing which all acids have in common, i.e. they all contain hydrogen which can be replaced by a metal to form a salt. We shall have more to say about salts later in the chapter. For example:

Magnesium + hydrochloric acid ⟶ magnesium chloride + hydrogen

Zinc + sulphuric acid ⟶ zinc sulphate + hydrogen

↑ METALS	↑ ACIDS	↑ SALTS	↑ HYDROGEN

The modern idea of acids

A more up-to-date idea of an acid is to regard it as a substance which can provide *hydrogen ions* in aqueous solution. You will remember that an ion is formed when an electron is added to or taken from a neutral atom. The hydrogen atom has one proton and one electron (it is the only atom without neutrons), and when the electron is removed we are left with the hydrogen ion which is really just a proton.

Hydrogen atom – one electron ⟶ hydrogen ion

All acids can produce hydrogen ions.

Experiment on the neutralisation of hydrochloric acid with sodium hydroxide solution (Fig. 40)

I. *Measuring the sodium hydroxide solution*
Using a pipette filler, transfer 25 cm^3 of sodium hydroxide solution to a flask. Add a few drops of indicator.

II. *Filling the burette*
Wash the burette with water and then with a few cm^3 of the hydrochloric acid. Pour the acid through a funnel into the burette which is held in position by a stand. Fill the part below the tap by allowing some acid to run out and then note the burette reading, taking care to read from the bottom of the meniscus. (See Fig. 40)

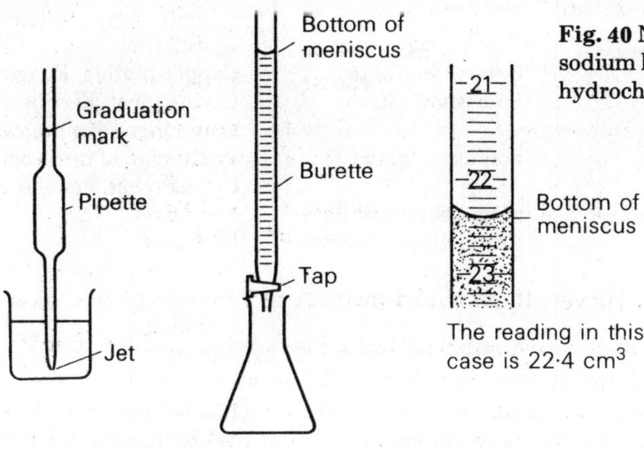

Fig. 40 Neutralisation of sodium hydroxide using hydrochloric acid

III. *The titration*

Allow the acid to run into the 25 cm^3 of sodium hydroxide solution and indicator so that about 2 cm^3 are added at a time, swirling the flask during each addition, and keeping the tap pressed in. When the red colour does not disappear almost immediately, run in the acid a drop at a time until the indicator remains pink. Note the burette reading, the difference between the two readings being the volume of acid needed to neutralise 25 cm^3 of sodium hydroxide solution. Then add this volume of fresh acid to a further 25 cm^3 of alkali without the indicator and evaporate the solution to dryness. If the solution starts "spitting" near the end, remove the burner.

Common salt

When hydrochloric acid is neutralised by sodium hydroxide, sodium chloride, common salt, is formed.

$$\text{Sodium hydroxide} + \text{hydrochloric acid} \longrightarrow \text{sodium chloride} + \text{water}$$

SALTS

When most people hear the word salt they probably think of ordinary table salt. This is common salt or sodium chloride and is just one of many salts used in everyday life.

Table of some important salts and their uses

Common name	Chemical name	Uses
Common salt	Sodium chloride	Production of sodium carbonate (washing soda), soap, sodium and chlorine
Blue vitriol	Copper sulphate	Copper plating, Fungicide
Saltpetre	Potassium nitrate	Curing meat. Fertiliser Component of gunpowder
Chile saltpetre	Sodium nitrate	Production of nitric acid Curing meat. Fertiliser
Epsom salts	Magnesium sulphate	Purgative
Sulphate of ammonia	Ammonium sulphate	Fertiliser
Washing soda	Sodium carbonate	Softening water, glass making
Bicarbonate of soda	Sodium hydrogen-carbonate	Baking
Hypo	Sodium thiosulphate	Photography
Chrome yellow	Lead chromate	Yellow paint

Salts are solids at normal temperatures and their crystals have well-defined shapes. Although salts vary in colour, crystalline shape, solubility in water, etc. they all have one thing in common. They are substances which are produced when the hydrogen ion of an acid is replaced by a metal ion or an ammonium ion NH_4^+.

Acid	Salts
Hydrochloric acid HCl	Chlorides, e.g. sodium chloride NaCl
Sulphuric acid H_2SO_4	Sulphates, e.g. copper sulphate $CuSO_4$
Nitric acid HNO_3	Nitrates, e.g. potassium nitrate KNO_3
Carbonic acid H_2CO_3	Carbonates, e.g. sodium carbonate Na_2CO_3

Acid salts

When only part of the hydrogen of an acid is replaced by a metal the product is called an *acid salt*.

Acid	Normal salt	Acid salt
Sulphuric acid H_2SO_4	Sodium sulphate Na_2SO_4	Sodium hydrogen sulphate $NaHSO_4$
Carbonic acid H_2CO_3	Sodium carbonate Na_2CO_3	Sodium hydrogencarbonate $NaHCO_3$

Only acids with more than one hydrogen atom per molecule can form acid salts.

How to make zinc sulphate crystals

Pour about 50 cm^3 of dilute sulphuric acid into a beaker and then add a few pieces of granulated zinc. If the reaction is slow, add a few drops of copper sulphate solution. Warm the mixture gently but do not boil. Note the evolution of hydrogen gas. If all the zinc dissolves, add more until excess zinc remains undissolved showing that all the acid has been used. Filter the mixture into an evaporating dish and boil the filtrate. Dip a glass rod into the solution, remove the rod and allow the solution on it to cool. If crystals appear on the rod, allow the whole of the solution to cool. Examine the crystals which separate, pour off the liquid and dry the crystals on a filter paper.

$$\text{Zinc} + \underset{\text{(dilute)}}{\text{sulphuric acid}} \longrightarrow \text{zinc sulphate} + \text{hydrogen}$$

How to make copper sulphate crystals

Copper sulphate crystals cannot be made as described above because copper will not displace hydrogen from dilute sulphuric acid.

However, copper oxide can be used. Warm, but do not boil, about 50 cm^3 of dilute sulphuric acid in a beaker. Add black copper oxide, a little at a time, stirring well after each addition, until no more will dissolve in the acid as shown by the presence of a black layer of oxide on the bottom of the beaker. Filter off the excess solid and treat the blue filtrate in an evaporating dish, as described above, to obtain dry crystals of copper sulphate.

Copper oxide + sulphuric acid ⟶ copper sulphate + water

Keep some of your crystals for later experiments.

How to make magnesium sulphate crystals

Magnesium sulphate crystals can be made by either of the two methods described above, using magnesium metal or magnesium oxide with dilute sulphuric acid.

Another method uses magnesium carbonate. Add small portions of magnesium carbonate to about 50 cm^3 of dilute sulphuric acid in a beaker until no further effervescence occurs showing that the acid has been neutralised. (What gas is evolved?) Filter, evaporate and obtain dry crystals as described above.

$$\begin{matrix}\text{Magnesium}\\ \text{carbonate}\end{matrix} + \begin{matrix}\text{sulphuric}\\ \text{acid}\end{matrix} \longrightarrow \begin{matrix}\text{magnesium}\\ \text{sulphate}\end{matrix} + \text{water} + \begin{matrix}\text{carbon}\\ \text{dioxide}\end{matrix}$$

How to make potassium nitrate crystals

Put 25 cm^3 of potassium hydroxide solution into a flask by means of a pipette filler. Add a few drops of an indicator, for example litmus, to the solution. Pour nitric acid solution through a funnel into a burette and allow a little of the acid to run through the burette tap to fill it. Note the burette reading and allow the acid to run into the potassium hydroxide solution a little at a time, shaking the flask after each addition. When the indicator *just changes* colour do not add any more acid as the alkali will have now been neutralised. Note the burette reading. The difference between the two readings will give the volume of nitric acid solution needed to neutralise 25 cm^3 of the potassium hydroxide solution. (See pages 85 and 86)

$$\begin{matrix}\text{Potassium}\\ \text{hydroxide}\end{matrix} + \begin{matrix}\text{nitric}\\ \text{acid}\end{matrix} \longrightarrow \begin{matrix}\text{potassium}\\ \text{nitrate}\end{matrix} + \text{water}$$

If the solution were evaporated the potassium nitrate would be coloured with indicator. Therefore, add the volume of acid used in the above reaction to a further 25 cm^3 of potassium hydroxide solution without the indicator. Transfer the solution to an evaporating dish, evaporate and obtain dry crystals as described in previous experiments.

How to make an insoluble salt – lead sulphate

Insoluble salts can be made by a *precipitation* reaction, e.g. in the case of lead sulphate, a soluble lead salt is dissolved in water and added to a solution of a soluble sulphate. As soon as lead and sulphate ions meet in solution a white precipitate of lead sulphate is produced.

$$Pb^{2+}\,(aq) + SO_4^{2-}\,(aq) \longrightarrow Pb^{2+}\,SO_4^{2-}\,(s)$$

Pb^{2+} (aq) means lead ions dissolved in water.
SO_4^{2-} (aq) means sulphate ions dissolved in water.
$PbSO_4$ (s) means lead sulphate solid in this case precipitated from solution because it is insoluble in water.

Dissolve some lead nitrate in water and add the solution to about 50 cm^3 of dilute sulphuric acid in a beaker. Lead sulphate is thrown down from solution as a white precipitate.

$$\text{Lead nitrate} + \text{sulphuric acid} \longrightarrow \text{lead sulphate} + \text{nitric acid}$$

Filter off the white precipitate, lead sulphate, and wash it with water to ensure it is not contaminated with any substance from the solutions. Dry the precipitate either in a steam oven or by heating the filter paper on a clock glass placed on a beaker of boiling water.

Soluble and insoluble salts

There is often difficulty in remembering which salts are insoluble. The following table should help:

Soluble		**Insoluble**
All common nitrates		
All common chlorides	**except**	Silver, lead and mercury (I) chlorides
All common sulphates	**except**	Lead, barium and calcium sulphates
Sodium, potassium and ammonium carbonates ←		All common carbonates **except**

Methods of making salts

The last five experiments illustrate five of the methods of salt making. We can now summarise them:

1. Action of a dilute acid on a metal.
2. Action of a dilute acid on a metal oxide.
3. Action of a dilute acid on a carbonate.
4. Action of a dilute acid on an alkali (for sodium, potassium and ammonium salts).
5. Precipitation – used for insoluble salts.

Water of crystallisation

Many crystals which have been obtained by crystallisation from aqueous solutions are found to contain water. The crystals are perfectly dry and the water is chemically combined within the crystals. This chemically combined water is called *water of crystallisation*.

A salt which contains water of crystallisation is said to be *hydrated* and a salt without it is said to be *anhydrous*. For example, copper sulphate:

$$\underset{\text{(blue)}}{\text{Hydrated copper sulphate}} \xrightarrow{\text{heat}} \underset{\text{(white)}}{\text{anhydrous copper sulphate}} + \text{water}$$

Anhydrous copper sulphate will turn blue in the presence of water and can be used to test for it.

To find out whether a salt has water of crystallisation present in it

Use the following salts:

1. Magnesium sulphate
2. Potassium nitrate
3. Washing soda
4. Sodium carbonate
5. Common salt
6. "Hypo" – sodium thiosulphate.

Heat each of these salts in a dry test-tube held horizontally and put a little anhydrous copper sulphate just inside the mouth of the tube (Fig. 41). Note any change and say whether you think the salts used have any water of crystallisation in them.

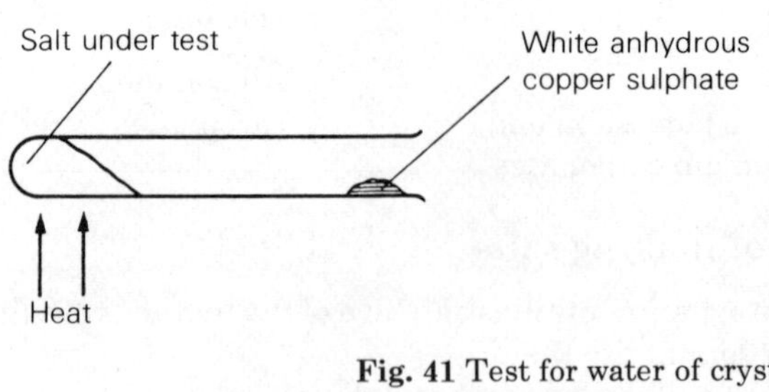

Fig. 41 Test for water of crystallisation

Efflorescence – a simple test with washing soda

Put some clear crystals of washing soda on a clock glass and leave them for a day or two. (The crystals must be clear. If they are not, remove any white powder from the surface before putting them on the clock glass.)

What do you see has happened after a few days?

On exposure to air, a white powder forms on the surface of the crystals. The reason is that washing soda loses some of its water of crystallisation to the air and the product is a white solid.

$$\text{Washing soda} \longrightarrow \underset{\text{(white powder)}}{\text{Sodium carbonate} - 1 - \text{water}} + \underset{\text{(lost to the air)}}{\text{water}}$$

The process of losing water of crystallisation to the air without heating is called *efflorescence.* Substances which do this are called *efflorescent* substances.

Deliquescence – a simple test on calcium chloride

On exposure to the air, anhydrous calcium chloride becomes sticky and slowly disappears leaving a clear liquid. The reason is that the salt absorbs water vapour from the air and dissolves in it to form a clear, aqueous solution.

Solids which have this property are said to be *deliquescent* and the process is called *deliquescence.* This property is not confined to salts; for example, sodium hydroxide, a base, is also deliquescent.

Most nitrates are deliquescent; potassium nitrate is one of the few non-deliquescent nitrates; if it were deliquescent it would be useless in gunpowder. Sodium chloride is not deliquescent and does not absorb water vapour from the air at all. The "clogging" of ordinary table salt would appear to contradict this but of course this is not pure sodium chloride. The impurity, magnesium chloride, being deliquescent, causes the "clogging".

Hygroscopic substances

Measure 25 cm^3 of concentrated sulphuric acid in a measuring cylinder and pour it into a beaker (care!) Leave it for a few days and then measure the volume of the liquid. Is it still 25 cm^3?

Solids which take water from the air and dissolve in it are called deliquescent substances. Some solids, e.g. copper oxide, take water from the air but do not dissolve in it to form solutions. Some liquids, e.g. concentrated sulphuric acid, also take water from the air. Such substances are said to be *hygroscopic.*

Uses of deliquescent and hygroscopic substances

The chief uses are:

1. In the desiccator
2. For drying gases

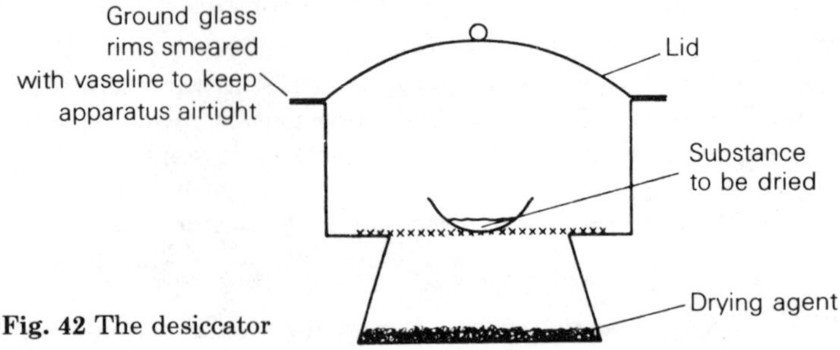

Fig. 42 The desiccator

The desiccator

The substance to be dried is placed in a suitable container on a metal gauze. Anhydrous calcium chloride or concentrated sulphuric acid is present in a container beneath the gauze.

Drying gases

The gas is led through concentrated sulphuric acid, anhydrous calcium chloride, quicklime, etc. the actual drying agent depending on the nature of the gas.

To find out the percentage by mass of water of crystallisation in barium chloride crystals

Weigh a clean, dry crucible and lid. One-third fill the crucible with powdered barium chloride crystals and weigh the crucible, lid and crystals. Heat the crucible on a pipe clay triangle, carefully at first and more strongly afterwards (Fig. 43). Leave a small space between the lid and the crucible so that water vapour can escape during heating. After heating for about 10 min allow the crucible, lid and contents to cool, preferably in a desiccator, then weigh again. Re-heat, cool and re-weigh to ensure that all the water has evaporated. Repeat the procedure until a constant mass is attained.

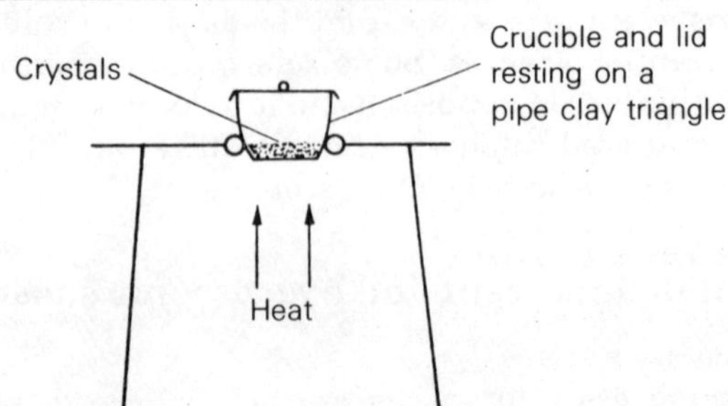

Fig. 43

Specimen calculation

Mass of crucible and lid	7·67 g
Mass of crucible and lid and crystals	9·77 g
Mass of crucible and lid and anhydrous salt I	9·46 g
II	9·46 g
Mass of crystals	2·10 g
Mass of water of crystallisation	0·31 g
Percentage by mass of water of crystallisation	$\frac{0{\cdot}31}{2{\cdot}10} \times 100$
	$= 14{\cdot}7$ per cent

To grow a large crystal of copper sulphate

Heat about 100 g of copper sulphate in about 200 cm^3 of water in a beaker. When the solid has dissolved filter the liquid if cloudy. Allow the clear liquid to cool slowly and when crystals have deposited pick out one with a good shape and tie it to one end of a length of cotton (use a crystal from your earlier preparation if you wish). Remove all the large crystals from the liquid and warm it just sufficiently to dissolve the small ones. Cool the liquid and suspend the chosen crystal in the solution (Fig. 44). Cover the beaker with paper and leave it in a cupboard for a day. The crystal will have grown. Remove it from solution, add a few more small crystals, warm to dissolve them, cool and suspend the crystal as before.

This procedure can be repeated several times to obtain a large crystal.

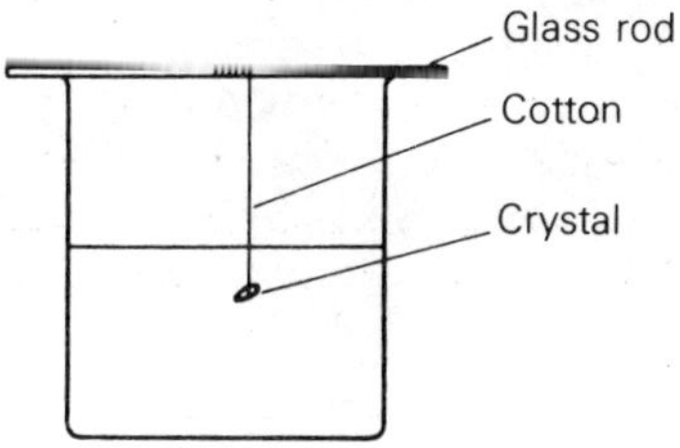

Fig. 44 Growing a crystal

Shapes of crystals

You will have noticed that the crystals of the different salts you have made do not all have the same shape.

In fact, there are altogether seven basic shapes of crystals with some variations (Fig. 45).

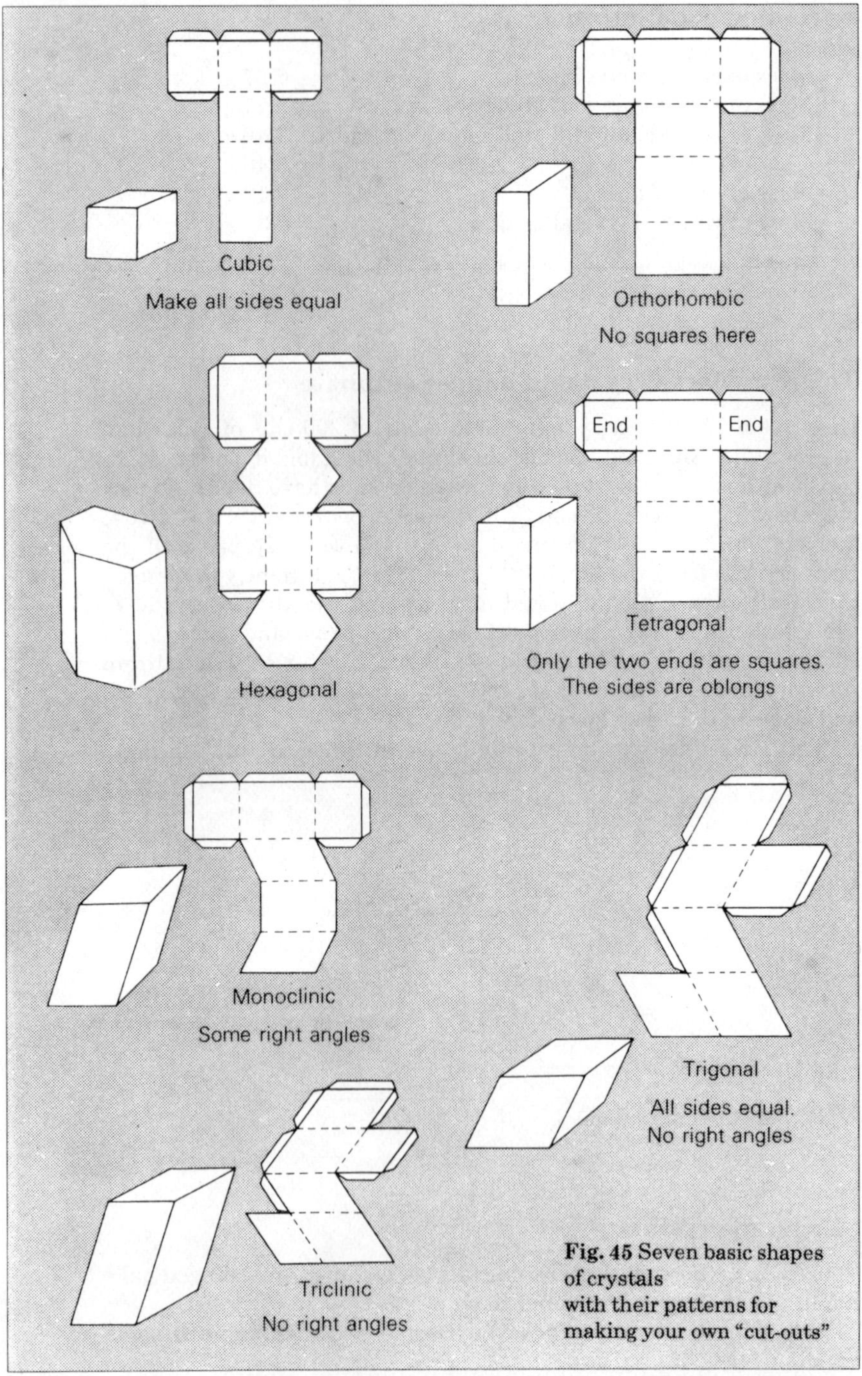

Fig. 45 Seven basic shapes of crystals with their patterns for making your own "cut-outs"

It is sometimes difficult to imagine what these shapes really look like in three dimensions. Perhaps you would like to make models of the seven basic patterns using some stout paper or cardboard. Make some patterns as shown in Fig. 45 and after cutting out, fold along the dotted lines and build up the models.

Fig. 46 Two variations of the cubic pattern. Can you see how these shapes can be obtained from a cube?

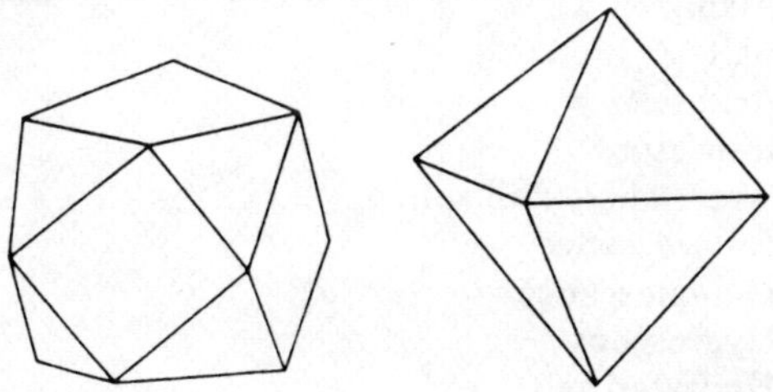

There are variations on each of the seven basic patterns. For example two variations of the cubic pattern are shown in Fig. 46.

Now have another look at the crystals of salts you have made and try to name the pattern to which each salt conforms.

Not only salts form crystals; metals can form really striking crystals.

Silver crystals

Wind a copper wire, which has been thoroughly cleaned and freed from grease, round a test-tube to make a spiral. Hook one end over the side of a beaker so that the spiral is suspended in a beaker of silver nitrate solution (Fig. 47). Leave for a day or two and note what happens.

Fig. 47 Growing silver crystals

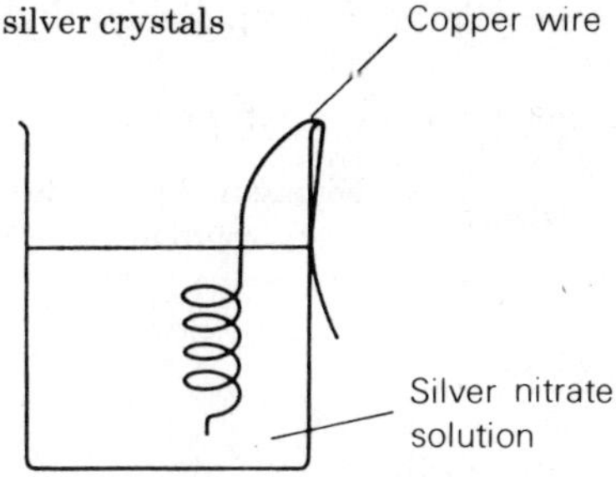

Lead crystals

Suspend a piece of granulated zinc in lead nitrate solution and inspect it after a few days.

N.B. If a projector is available it is possible to immerse a metal in a gel of a suitable salt and actually see crystals growing.

THINGS TO REMEMBER

You should know now what is meant by:

- Acid
- Base
- Alkali
- Salt
- Acid salt
- Indicator
- Water of crystallisation
- Efflorescence
- Deliquescence
- Hygroscopy
- Hydrated salt
- Anhydrous salt.

QUESTIONS TO TRY

1. Why do some fruits have a sour taste?

2. Name the three common mineral acids.

3. Suppose you were given three bottles labelled *A*, *B* and *C* and were told that one contained hydrochloric acid, another sodium hydroxide solution and the third water. How would you attempt to identify the three liquids using methyl orange and phenolphthalein.

4. Complete the following giving the appropriate colour changes:

	Nitric acid	Potassium hydroxide solution	Sulphuric acid
Methyl orange			
Litmus			
Phenolphthalein			

5. Describe how you would obtain:
(a) potassium sulphate crystals from sulphuric acid;
(b) sodium nitrate crystals from sodium hydroxide;
(c) potassium chloride crystals from hydrochloric acid.

6. Complete the following:
(a) ... hydroxide + nitric acid $\longrightarrow$ ammonium ... + water.
(b) Sodium hydroxide + ... acid $\longrightarrow$ ... sulphate + water.
(c) ... hydroxide + ... acid $\longrightarrow$ potassium chloride + water.

7. Suppose you were given *six* solutions *A, B, C, D, E* and *F* and the pH scales of these solutions were respectively 2, 3, 4, 5, 8 and 10.
   (a) Which solution is most acidic?
   (b) Which solution is most alkaline?
   (c) What would be the effect on the pH of *B* of adding lime to it?
   (d) What would be the effect on the pH of *E* of adding vinegar to it?
   (e) If you were given a solution of universal indicator but not its colour chart, describe how you would use these six solutions to help you find the pH of lemon juice.

8. Describe how you could prepare:
   (a) Copper sulphate from copper carbonate.
   (b) Magnesium sulphate from magnesium oxide.
   (c) Barium sulphate from barium chloride.

9. If you were given sodium hydroxide, ammonium hydroxide, nitric acid and hydrochloric acid how many salts could you prepare by neutralisation? Give their names.

10. Write word equations for the neutralisation of:
    (a) sodium hydroxide by nitric acid;
    (b) ammonium hydroxide by sulphuric acid;
    (c) potassium hydroxide by sulphuric acid.

11. Write word equations for the neutralisation of:
    (a) hydrochloric acid;
    (b) sulphuric acid;
    (c) nitric acid;

    by copper oxide in each case.

12. Why cannot sodium nitrate be used in place of potassium nitrate in gunpowder although weight for weight it provides more oxygen?

13. The following results were obtained in an experiment with magnesium sulphate crystals.

Mass of crucible	= 14·636 g
Mass of crucible + crystals	= 15·374 g
Mass of crucible and anhydrous salt	= 14·996 g

    What is the percentage of water of crystallisation in the crystals?

14. Citric acid is present in both oranges and lemons, yet oranges are sweet and lemons are sour. Can you explain this?

7 Metals

Man uses the word "metal" in work or in the home without really thinking what it means. Metals are so much an essential part of our everyday lives that it is easy to give examples of their very wide use. Copper is used for piping domestic water, in central heating systems and also for electrical wiring; iron, usually in the form of steel, is used for motor vehicles and in prestressed concrete for building purposes; silver and gold are used for jewellery. It is much easier to give examples of the use of metals than to give a precise definition of what a metal actually is.

Scientists say that metals have several definite properties:

1. They conduct electricity well.
   This is because they have electrons which can be taken out at one end of a piece of metal and pushed in at the other, rather like water in a pipe.
2. They can be made into wires and sheets by drawing and hammering. As a result, they are said to be malleable and ductile.
3. They ring when struck as long as they are placed somewhere where they can vibrate.
4. When they have no coating of oxide they are shiny. Even dull metals can be polished, but they often become dull again very quickly.
5. Many metals can be made to melt between 500°C and 1500°C, and can be made into tools, parts of machines, building materials and very many other things. Most types of metals can be joined together very strongly by soldering, welding or by the use of a special adhesive.

Our society is based on the use of metals which when dug from the earth will often look very different from the pure metals because of their combination with other elements. Aluminium, the most abundant metal on earth, occurring in clays and oxides, is very different in appearance from these ores. Furthermore, aluminium is not easily extracted and it was not until 1845 that the pure metal was isolated. The process was very expensive, making aluminium a valuable metal. Aluminium cutlery was a status symbol in those days and it was not until 1886 that a relatively cheap process of extracting aluminium was invented. Titanium, as a commercial metal, is even newer. Until 1955 it was a laboratory curiosity but now several tonnes of this light, strong metal are used in Concorde and other aircraft of today.

Even when metals have been recovered from their ores many are useless as they stand for manufacturing goods. Some are too soft; some may crack easily; some may melt at too low a temperature or corrode too rapidly. In order to overcome such problems it is necessary to mix together quantities of different metals to achieve the properties required.

These mixtures of metals are called alloys.

Examples	Bronze	=	copper + tin
	Brass	=	copper + zinc
	Steel	=	iron + carbon + other elements

Steel is probably the best known and most widely used alloy. By varying the small proportions of other elements mixed with iron and carbon, a wide range of different steels can be made. The use of chromium along with iron and carbon will, for example, produce stainless steel.

The study of metals, alloys, their properties and uses is called metallurgy.

Components, such as gears, need to be relatively soft for machining yet hard-wearing in use. Thin metal foils on milk bottles, North Sea oil rigs, ring-pull drink cans, and dental fillings illustrate the breadth and depth of the study of metallurgy.

Nuclear power, aerospace and off-shore engineering indicate the scale of the future for metallurgy.

Some of the more common metals

Although there are 78 elements that are generally recognised as being metals, comparatively few of them are used by engineers to make components for manufactured goods. Some are dangerously radioactive, some are very, very expensive, far more expensive than gold, and many have properties that make them less useful than cheaper and more common metals.

When we carry out experiments to find out about the properties of metals we usually study only twelve of them.

1. *Potassium.* This is a very reactive metal which reacts violently with water, cannot be touched with the skin, and has to be stored under oil. It is quite soft, can be cut by a knife exposing a shiny new surface which soon tarnishes. The metal floats on and reacts vigorously with water.

2. *Sodium.* This has similar properties to potassium, although it is not quite as reactive. It is, however, still very dangerous to handle.
3. *Calcium.* It is normally sold in the form of small pellets which are often stored under oil. Calcium is denser than water and also reacts with it.

These three metals are not used to make tools or engineering components because they are too soft and too reactive.

4. *Magnesium.* It is a metal of low density (1·74 g/cm^3) and tarnishes easily.
5. *Aluminium.* This also has a low density (2·70 g/cm^3) and has a bright, shiny appearance.
6. *Zinc.* This is dull, although it is bright when polished. Its density is 7·13 g/cm^3.
7. *Iron.* It is shiny when polished, but rusts easily and is moderately dense (7·86 g/cm^3).
8. *Lead.* This metal is very dense (11·35 g/cm^3) and rapidly becomes coated with a dull layer in air. Both the metal and its compounds are poisonous.
9. *Copper.* When pure it is an excellent conductor of electricity. It is a shiny, orange metal of density 8·94 g/cm^3.
10. *Mercury.* An unusual metal in that it is a liquid at room temperature. When it is frozen solid it behaves rather like lead. The density of the liquid is 13·53 g/cm^3.
11. *Silver.* } Both are dense metals with densities of 10·50 g/cm^3 and
12. *Gold.* } 19·32 g/cm^3 respectively. Both are expensive and used in the manufacture of valuable objects and in industry where metals that do not tarnish easily are needed.

THE REACTIVITY TABLE OF METALS

In Chapter 3 you used the idea of an acid and an alkali. An acid is a substance which turns litmus red, or universal indicator to a colour indicating a pH of less than 7.

An alkali is a substance which turns litmus blue and turns universal indicator to a colour indicating a pH of more than 7. When metals burn in air and their oxides react with water and dissolve, they produce an alkaline solution.

$$\text{Metal oxide} + \text{water} \longrightarrow \text{metal hydroxide (an alkali)}$$

$$\text{e.g. Calcium oxide} + \text{water} \longrightarrow \text{calcium hydroxide}$$

We can make a list of some metals and non-metals in the order of their activity with oxygen.

Potassium
Sodium
Calcium
Magnesium
Aluminium
Carbon
Zinc
Hydrogen
Iron
Lead
Copper
Mercury
Silver
Gold

If some of the metals low down in the list are heated, they can be made to burn, but they need to be in the form of fine wire or powder.

How does the state of a metal affect its burning in air?

Heat some zinc powder on an old, clean milk bottle-top, holding the milk bottle-top well above the top of the flame with some tongs, so that it does not melt. If the zinc powder is gently shaken it burns readily, producing a powder that floats in air.

Zinc + oxygen ⟶ zinc oxide (yellow when hot; white when cold)

Now try heating zinc foil and compare the result with the above.

Steel (iron) wool will burn when held by tongs in a Bunsen flame producing sparks of burning iron. Now try heating a steel nail in tongs. Is there any difference?

The state of the metal, powder, wire or sheet tends to affect the rate of oxidation, as well as the nature of the metal.

METALS COMPETING FOR OXYGEN

The reactive metals tend to hold on to oxygen in their oxides and do not let go of it easily. The unreactive metals on the other hand can be made to give up the oxygen from their oxides quite easily. Silver oxide can be decomposed to silver and oxygen by strong heating, and gold does not even form an oxide!

We can make metals "compete" for oxygen by heating a metal and a metal oxide together.

Competition for oxygen

Heat the following mixtures in turn in a crucible, using a pipe-clay triangle and tripod.

(a) Iron filings + zinc oxide
(b) Iron filings + copper oxide
(c) Iron filings + lead (II) oxide

Do not put the lid on the crucible, but wear goggles and look for a glow that indicates some reaction has occurred in the crucible. When the mixture has cooled, examine to see if any change has occurred.

Note that lead (II) oxide is poisonous. You *must* wash your hands at the end of the experiment.

If a metal oxide has its oxygen taken away from it, we say that it has been *reduced.* Reduction is the opposite to oxidation.

If a metal oxide has been reduced, then this suggests that you have been heating it with a more active metal. The metals "fight" for oxygen rather like a big boy and a small boy having a tug-of-war for a ball. The big boy always wins and the more active metal always wins.

It is possible to make a sample of iron by reducing some iron oxide with aluminium powder in a fire-clay crucible. You must *not* try this reaction yourself as it is very dangerous. It is known as the *thermite reaction.*

$$\text{Aluminium} + \text{iron oxide} \longrightarrow \text{iron} + \text{aluminium oxide}$$

Carbon and its place in the series

Just as you heated iron with some metal oxides, you can heat charcoal (the element carbon) with some metal oxides.

Try heating the following mixtures in a crucible *with the lid on* and look for the metal after heating for 5 minutes.

(i) Magnesium oxide + carbon
(ii) Lead (II) oxide + carbon
Remember lead (II) oxide is poisonous.

You will find out if the reactivity table fits your results for these experiments.

THE REACTION OF METALS WITH WATER

Potassium and sodium are very reactive metals and each readily attacks water to give hydrogen gas and an alkaline solution. As the reactions are very vigorous they will probably be demonstrated by the teacher (see Chapter 8, pages 111–12).

You can, however, try the reactions of some other metals with water. Obtain small pieces of the following metals:

Zinc, magnesium, copper, aluminium, iron calcium, lead.

Investigate the reactivity of each metal with water using the apparatus shown (Fig. 48).

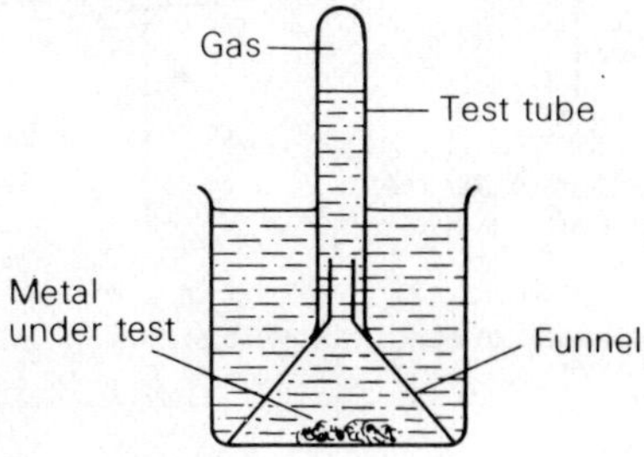

Fig. 48 Apparatus to test the reactivity of metal with water

Make sure that you use cold, distilled water and determine the pH of the water by adding a few drops of universal indicator solution before you start the experiment. Leave the apparatus to stand and after 15 minutes observe carefully to see if any bubbles of gas are formed. If sufficient gas to test has been evolved, invert the tube, holding your finger over the end, and then test the gas in the Bunsen flame for hydrogen. Record any pH change which occurs.

THE REACTION OF METALS WITH STEAM

Although some metals show little, if any, reaction with water at room temperature they may react at a higher temperature.

Try the reaction of magnesium, zinc and iron with steam using the following apparatus (Fig. 49). You *must* use a safety screen between yourself and the experiment and you *must* wear goggles.

Use hard glass (pyrex) test-tubes and carry out three separate experiments in turn.

1. Take 20 cm of magnesium ribbon coiled round a pencil. This experiment ruins the tube so your teacher may prefer to demonstrate it.

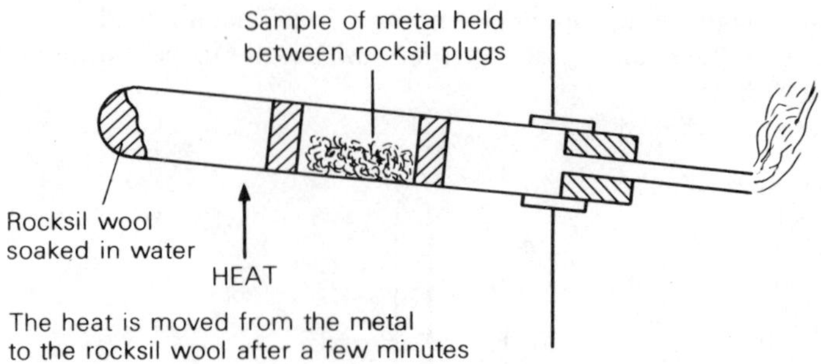

Fig. 49 Testing the reaction of a metal with steam

2. A piece of zinc foil 6 cm x 12 cm rolled to fit the tube.
3. Iron filings between two Rocksil plugs as illustrated.

Try to set light to the gas being evolved from the tube.

Not all the reactions will be seen to fit into our ideas about the activity table. Aluminium, for instance, is a metal that has an invisible, protective coating of aluminium oxide, and so it does not seem to react easily, even though it comes high in the table. The metals after iron in the table seem to be unable to react with water.

If the metals react, they do so:

$$\text{Metal} + \underset{\text{(water)}}{\text{hydrogen oxide}} \longrightarrow \text{hydrogen} + \text{metal oxide}$$

With potassium, sodium and calcium the metal hydroxide is produced instead of the oxide.

The reactivity table and the reaction of metals with dilute hydrochloric acid

When we made hydrogen we made use of the reaction of zinc with dilute acid. If we wish to establish that the activity series we have written agrees with observations made on this reaction we can do the following experiment.

Place a small piece of each of the following metals in a test-tube, and add 2–3 cm^3 of dilute hydrochloric acid after putting on goggles. Warm, if necessary, and then test any gas produced for hydrogen by placing an inverted boiling tube over the mouth of the test-tube and then after one minute, test with a Bunsen flame. Copy out the table and fill it in.

Metal	Hydrogen gas produced *Yes or No*
Zinc	Yes
Aluminium	
Iron	
Lead	
Copper	
Magnesium	

Do the results of your class and yourself support the idea of the activity table?
Write word equations for each recorded reaction, and make a note of metals which do *not* react with dilute acid.

THE PRODUCTION OF IRON

We have seen that carbon can reduce many metal oxides to the metal. From the reactivity table it can be seen that iron oxide can be reduced by carbon. Carbon in the form of charcoal or coke has always been plentiful and cheap, and so it has been used for iron making.

There are several ores of iron. Magnetite is naturally magnetic and the first compass was made from it. Haematite has a red colour which it often lends to the earth in the areas where it is found. These ores are different oxides of iron and can be reduced by carbon.

In many parts of the world after about 2000 B.C. men learned to heat iron ore in a charcoal fire and make a mass of soft iron which could be hammered and shaped and made into weapons and tools. During the Middle Ages, iron-making furnaces were improved by passing a blast of air from bellows into the furnace to make it hotter. So much forest was cut down in the seventeenth century that coke, produced by heating coal in the absence of air, replaced charcoal as a source of carbon for furnaces.

The blast-furnace (Fig. 50)

The huge amounts of iron produced today are made in blast-furnaces.

A blast of hot air is fed into the furnace which contains iron oxide (ore) and carbon (coke). At temperatures of over 1100°C reduction of the iron oxide occurs.

$$\text{Iron oxide} + \text{carbon} \longrightarrow \text{iron} + \text{carbon dioxide}$$

Actually, the process takes place in several more complicated steps:

(a) At the bottom of the furnace:

Iron oxide + carbon ⟶ iron + carbon monoxide

(b) Higher up the furnace:

Iron oxide + carbon monoxide ⟶ iron + carbon dioxide

In order to remove some of the impurities from the molten iron, limestone, a form of calcium carbonate, is added to the materials in the furnace. This removes sand and other impurities by combining with them to make a slag which floats on top of the molten iron.

Fig. 50 The blast-furnace

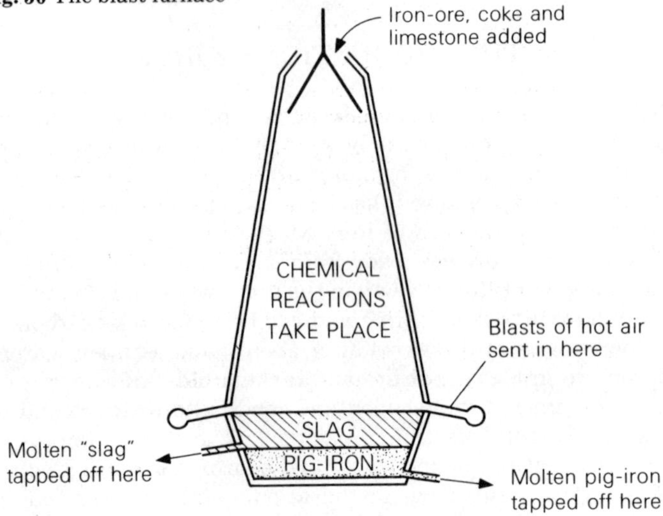

The iron produced by the blast-furnace is called cast iron or pig-iron and contains up to 7 per cent of impurities such as carbon, silicon, phosphorus, manganese and sulphur. Cast iron is very brittle. Pure iron is called wrought iron and is very soft.

Steel contains just a small amount of impurity and is very hard and tough but not brittle. Steels are alloys of iron with other metals. Steel can be made by blowing pure oxygen over the surface of molten, cast iron in a special vessel called a converter. The impurities are oxidised and some escape as gaseous oxides, until just the correct small percentages remain. This method of making steel is called the LD process.

Production of other metals

All metals up to zinc in the table can be produced by reducing the oxide with carbon, but the metals above zinc can only be made by taking a molten compound of the metal, and passing electricity through it. The metal is formed at the cathode. The most important metal produced in this way is aluminium.

Table of some uses of metals

Sodium and Potassium	Both have low melting points and are used as coolants in nuclear reactors.
Magnesium	Mixed with aluminium and copper it produces a strong, light alloy used in aircraft manufacture.
Aluminium	Saucepans; cooking foil; car engines; ship superstructures; all are made from this metal.
Zinc	Galvanising – a method of coating iron with zinc to prevent rusting.
Iron	Used for making different steels: also tools, cars, construction scaffolding and framework.
Lead	Car batteries; type-metal alloy (86 per cent lead, 11 per cent antimony, 3 per cent tin). Solders (Lead m.p. 327°C. Tin m.p. 232°C. Alloys of lead and tin have even lower melting points.)
Copper	Commonly used for electrical wiring and water pipes.
Mercury	Thermometers have mercury in the bulb at the base of the glass tube up which it rises to record temperatures.
Silver and Gold	Both these precious metals are used in jewellery and electrical equipment because they do not tarnish rapidly.

8 The Periodic Table

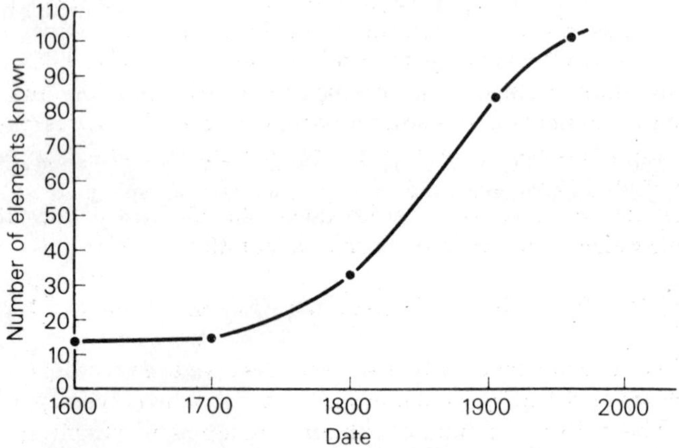

Fig. 51 Graph showing the increase in the number of known elements

Sometime during your life you will no doubt have been told to put some of your belongings into their proper place. You may have left some books scattered about, or some clothes, or possibly some sports equipment. The request may have been a little unpopular with you at the time but if you think about it, it is very convenient in many ways to have certain places for certain things – in other words, to create order out of disorder.

As elements were discovered and the number of known elements increased (Fig. 51), chemists tried to put them into some sort of order, or classify them as we say. Lavoisier, notable for his researches into the composition of air, was one of the first to attempt this. In his day only about thirty elements were known and his classification consisted chiefly of dividing the elements into metals and non-metals.

Relative atomic mass

It was realised in the nineteenth century that atoms of different elements have different masses. Hydrogen atoms are the lightest of all; one million million million million of them have a mass of only a gram, so you will see that it is very inconvenient to talk about the mass of one single hydrogen atom. It would involve too many noughts. To simplify the picture, hydrogen was said to have a mass of *one* unit or to have a relative atomic mass of 1, and as the atoms of all the other elements were heavier than the hydrogen atom they were assigned relative atomic masses greater than 1. One sodium atom is 23 times as heavy as a hydrogen atom and was therefore given a relative atomic mass of 23. One iodine atom is 127 times as heavy as a hydrogen atom and was therefore given a relative atomic mass of 127, etc., etc.

Newlands in 1864, in an attempt to classify the elements, wrote down the elements in the order of increasing atomic mass and noticed that any particular element was very similar to the eighth element away from it in the list.

Let us, for illustration, write down the elements from lithium, atomic weight 7, to calcium, atomic weight 40.

(Li) Be B C N O F (Na) Mg Al Si P S Cl (K) Ca

If we consider the element sodium (Na), we see that the elements which are eighth away from it in the list are lithium (Li) and potassium (K). These three elements, lithium, sodium and potassium, have very similar properties and form a "chemical family". If you study the list given above you will see that beryllium (Be), magnesium (Mg) and calcium (Ca) should form another chemical family of similar elements. Newlands found that this was the case.

Newlands' attempt at classifying the elements was named "The Law of Octaves" and at the time was greeted with a certain amount of ridicule. Some people thought that it was sheer luck that there was some relationship between atomic mass and properties, and one or two people sarcastically suggested that Newlands should arrange the elements in alphabetical order.

Mendeleef's Periodic Table

Not long after Newlands's attempt at classifying elements, the Russian chemist, Dmitri Ivanovich Mendeleef, showed that octaves alone would not be sufficient to provide a completely satisfactory classification. In some cases a repetition of properties appeared after *eighteen* elements, not eight, and even then there were a number of other snags. However, like Newlands, Mendeleef arranged the elements in the order of increasing atomic mass, but placed them in a table (Fig. 52), called the Periodic Table because the same properties occur in the elements at certain periods.

It is worth having a look at Mendeleef's Table and noting one or two points. There are no noble gases included because they had not been discovered in his time. The dashes mark the gaps he left for undiscovered elements; it was very necessary to leave such gaps so that the then known elements would fall into the right family or *Group* of elements. He seemed to be perfectly justified in doing this because the pattern of properties was so remarkable. Elements which fell into any one particular vertical column or *Group*, as we say, have similar chemical properties.

Series	Group 1	Group 2	Group 3	Group 4	Group 5	Group 6	Group 7	Group 8
1	H							
2	Li	Be	B	C	N	O	F	
3	Na	Mg	Al	Si	P	S	Cl	
4	K	Ca	—	Ti	V	Cr	Mn	Fe Co Ni
5	Cu	Zn	—	—	As	Se	Br	
6	Rb	Sr	Y	Zr	Nb	Mo	—	Ru Rh Pd
7	Ag	Cd	In	Sn	Sb	Te	I	
8	Cs	Ba	Dy	Ce	—	—	—	
9	—	—	—	—	—	—	—	
10	—	—	Er	La	Ta	W	—	Os Ir Pt
11	Au	Hg	Tl	Pb	Bi	—	—	
12	—	—	—	Th	—	U		

Fig. 52 Mendeleef's Periodic Table

One difficulty that Mendeleef found in compiling the table was that in a few cases, two consecutive elements fell in the "wrong" groups but a straightforward switch of the two placed them in the "right" groups. For example, according to atomic masses, iodine should be in Group 6 and tellurium in Group 7, but in properties tellurium is like Group 6 elements and iodine is like Group 7 elements. Mendeleef simply did the switch and assumed an explanation would follow later.

The modern Periodic Table

Newlands and Mendeleef knew nothing about the structure of the atom and their attempts at classifying the elements were based entirely on atomic masses. Since the 1890s, however, when the electron was discovered, we have been able to build up a fairly good picture of the atom and we have already discussed this in Chapter 4. You will remember that the *atomic number* of an element is the number of electrons or protons in one atom of the element. Let us arrange the first twenty elements in the order of atomic number.

Hydrogen	H	1	Sodium	Na	11
Helium	He	2	Magnesium	Mg	12
Lithium	Li	3	Aluminium	Al	13
Beryllium	Be	4	Silicon	Si	14
Boron	B	5	Phosphorus	P	15
Carbon	C	6	Sulphur	S	16
Nitrogen	N	7	Chlorine	Cl	17
Oxygen	O	8	Argon	Ar	18
Fluorine	F	9	Potassium	K	19
Neon	Ne	10	Calcium	Ca	20

We can arrange these elements in much the same way as did Mendeleef, but of course he was not aware of the existence of the rare gases.

We thus have: (H)

							He
Li	Be	B	C	N	O	F	Ne
Na	Mg	Al	Si	P	S	Cl	Ar
K	Ca						

Let us now have a look at the properties of these elements.

1. *Sodium* Your teacher may clamp a hard glass tube in a vertical position so that its end just dips below the surface of water in a trough (Fig. 53). When a small piece of sodium is placed down the tube it will react with the water as soon as it touches it. Hydrogen gas liberated will burn at the mouth of the tube and the addition of an indicator will show that the solution formed in the trough is alkaline.

Sodium + water ⟶ sodium hydroxide + hydrogen

Fig. 53 Action of sodium on water

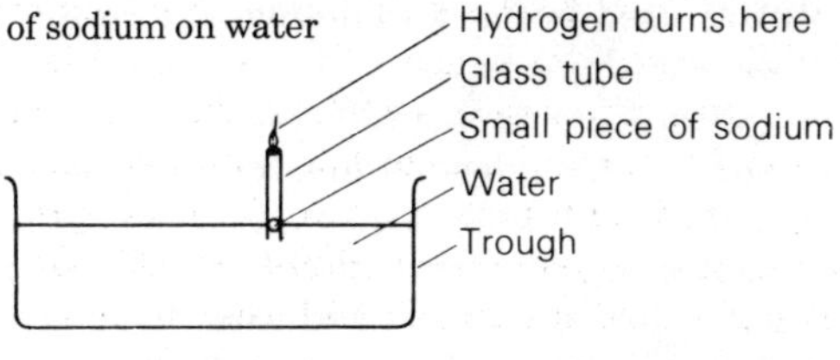

2. *Potassium* Your teacher may add a very small piece of potassium, about the size of a small pea, to a trough of water. You will notice that the potassium catches fire and floats rapidly along the surface of the water. Eventually the potassium will all disappear leaving an alkaline solution.

$$\text{Potassium} + \text{water} \longrightarrow \text{potassium hydroxide} + \text{hydrogen}$$

3. *Calcium* When a few calcium turnings are added to water in a beaker containing an inverted glass funnel and test tube as shown (Fig. 54) hydrogen gas is evolved. The addition of an indicator to the remaining solution will indicate alkalinity.

$$\text{Calcium} + \text{water} \longrightarrow \text{calcium hydroxide} + \text{hydrogen}$$

What test could you apply to the hydrogen gas collected?

Fig. 54 Action of calcium on water

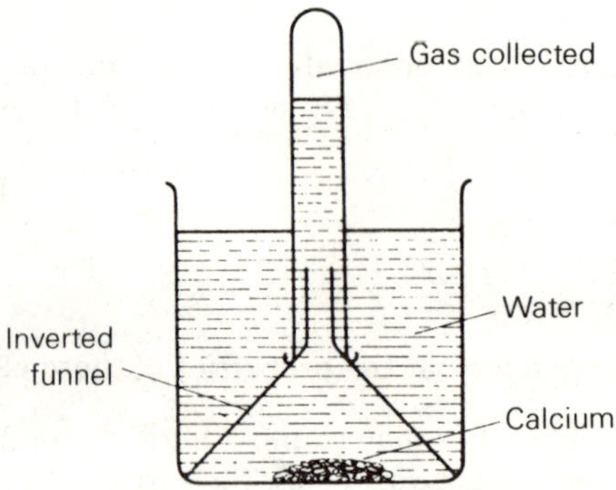

4. *Magnesium* Add a piece of magnesium ribbon to water in a beaker and note what happens. Then take another piece of magnesium ribbon and scrape off the outer coating with a pen-knife. Look at the magnesium. What is the difference, if any? Add this piece to water in the apparatus as used in Fig. 54 and leave it for a few days. Add pH indicator. Note what happens.

5. *Some non-metals.* Place some carbon and sulphur in water. Is there any reaction? Nitrogen and oxygen, two of the gases of the air, have no reaction with cold water, neither do helium, neon and argon.

Fluorine and chlorine *do* react with cold water but no hydrogen is liberated and in this way differ from the metals described above.

Chemical families or groups

GROUP 1. *The alkali metals*

Lithium, sodium and potassium fall into this family of elements. They are very similar indeed in their properties; for example, they are all easily oxidised by the air to form oxides and to prevent this happening are stored under oil. They are soft metals which can be cut with a knife very easily indeed, and they all have a valency of 1. They have a relative density of less than 1 and so float on water which reacts with them very vigorously to form hydrogen and an alkaline solution. For example:

$$\text{Sodium} + \text{water} \longrightarrow \text{sodium hydroxide} + \text{hydrogen}$$
$$2Na + 2H_2O \longrightarrow 2NaOH + H_2$$

In the experiments described above the formation of an alkaline solution during the reaction between water and these alkali metals is shown by the indicator.

GROUP 2. *The alkaline earth metals*

Beryllium, magnesium and calcium are members of this family and although they will react with water to form an alkaline solution with liberation of hydrogen they are not nearly as vigorous as the alkali metals in doing so. They also differ from the alkali metals in having a valency of 2, and they are all denser than water.

$$\text{Calcium} + \text{water} \longrightarrow \text{calcium hydroxide} + \text{hydrogen}$$
$$Ca + 2H_2O \longrightarrow Ca(OH)_2 + H_2$$

As far as the other elements up to calcium are concerned, we find that boron and aluminium show some resemblance, carbon and silicon are alike in some ways, nitrogen resembles phosphorus, oxygen is similar to sulphur, fluorine and chlorine behave very much alike and helium, neon and argon are very much alike. Elements in the outer columns of the table are more alike than those in the centre columns.

The modern Periodic Table is made up of *all* the elements arranged in the order of atomic numbers (Fig. 55). It may look very forbidding but do not worry, you will not be required to learn it. The table is one of the backbones of chemistry and you will find it more interesting and important as you learn how the properties of an element are governed by its place in the table.

You may have been wondering why elements should be arranged in the order of their atomic numbers and not their atomic masses. Actually there is very little difference in the two arrangements, the only difference, admittedly an important one, is found in cases where

Fig. 55 The Periodic Table

H Hydrogen 1																	He Helium 2
Li Lithium 3	Be Beryllium 4											B Boron 5	C Carbon 6	N Nitrogen 7	O Oxygen 8	F Fluorine 9	Ne Neon 10
Na Sodium 11	Mg Magnesium 12											Al Aluminium 13	Si Silicon 14	P Phosphorus 15	S Sulphur 16	Cl Chlorine 17	Ar Argon 18
K Potassium 19	Ca Calcium 20	Sc Scandium 21	Ti Titanium 22	V Vanadium 23	Cr Chromium 24	Mn Manganese 25	Fe Iron 26	Co Cobalt 27	Ni Nickel 28	Cu Copper 29	Zn Zinc 30	Ga Gallium 31	Ge Germanium 32	As Arsenic 33	Se Selenium 34	Br Bromine 35	Kr Krypton 36
Rb Rubidium 37	Sr Strontium 38	Y Yttrium 39	Zr Zirconium 40	Nb Niobium 41	Mo Molybdenum 42	Tc Technetium 43	Ru Ruthenium 44	Rh Rhodium 45	Pd Palladium 46	Ag Silver 47	Cd Cadmium 48	In Indium 49	Sn Tin 50	Sb Antimony 51	Te Tellurium 52	I Iodine 53	Xe Xenon 54
Cs Cæsium 55	Ba Barium 56	La Lanthanum 57	Hf Hafnium 72	Ta Tantalum 73	W Tungsten 74	Re Rhenium 75	Os Osmium 76	Ir Iridium 77	Pt Platinum 78	Au Gold 79	Hg Mercury 80	Tl Thallium 81	Pb Lead 82	Bi Bismuth 83	Po Polonium 84	At Astatine 85	Rn Radon 86
Fr Francium 87	Ra Radium 88	Ac Actinium 89															

La Lanthanum 57	Ce Cerium 58	Pr Praseodymium 59	Nd Neodymium 60	Pm Promethium 61	Sm Samarium 62	Eu Europium 63	Gd Gadolinium 64	Tb Terbium 65	Dy Dysprosium 66	Ho Holmium 67	Er Erbium 68	Tm Thulium 69	Yb Ytterbium 70	Lu Lutetium 71
Ac Actinium 89	Th Thorium 90	Pa Protoactinium 91	U Uranium 92	Np Neptunium 93	Pu Plutonium 94	Am Americium 95	Cm Curium 96	Bk Berkelium 97	Cf Californium 98	Es Einsteinium 99	Fm Fermium 100	Md Mendelevium 101	No Nobelium 102	Lw Lawrencium 103

two consecutive elements are wrongly placed when atomic masses are used but rightly placed when atomic numbers are used. When Mendeleef compiled his Table, he knew nothing about electrons, and therefore atomic numbers, so he could not explain the discrepancies with elements like tellurium and iodine. It is the number of electrons in an atom, not the atomic mass, which governs the properties of an element. Later in your chemistry lessons you will learn how the electrons are arranged inside atoms. Mendeleef was fortunate in that he got more or less the same arrangement of elements using atomic masses as we do today using atomic numbers.

Valency and the Periodic Table

In Chapter 4 we said that valencies must be learned and the list given may have looked a little forbidding and without system. The Periodic Table helps us to remember valencies.

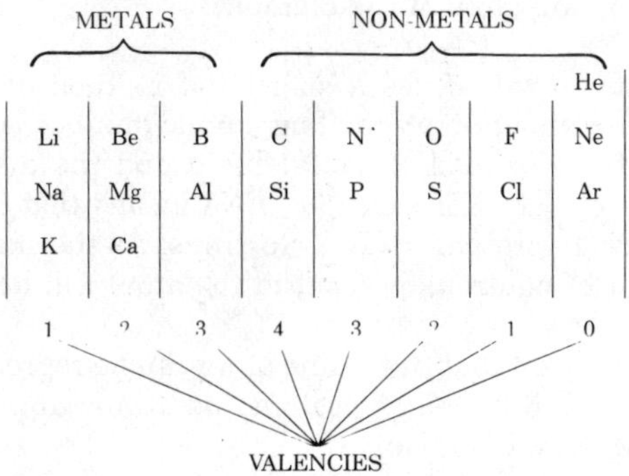

You can refer to the full Periodic Table and the same "rules" apply. The element barium, for example, has a valency of 2 because it is a member of Group 2. Iodine has a valency of 1 because it is a member of Group 7, like fluorine and chlorine. There are other factors which will emerge as you progress in your chemistry studies.

Metals and non-metals

As we look from left to right of the Periodic Table, for example, from sodium to chlorine we find that there is a gradual change from metal to non-metal. It follows from this that the oxides of elements to the left of the table will tend to be *basic* while oxides of elements to the right of the table will tend to be *acidic*. This fits in very nicely with what we found in Chapter 3. You will remember that sodium, calcium and magnesium produce basic oxides when burned in oxygen while carbon, sulphur and phosphorus produce acidic oxides.

The Periodic Table will really help us to bring some system into our chemistry.

ISOTOPES

Although the terms isotope, radioactivity, radioisotope and atomic energy are familiar enough, they are nevertheless comparatively new and many people alive today can remember when they were first introduced. Let us now have a look at each of these terms.

Isotopes

You will remember that

(i) The *atomic number* of an element is the number of protons or electrons in one atom of the element.

(ii) The *mass number* is the number of protons + the number of neutrons in one atom of the element.

We find it convenient to use a shorthand method of writing down this information about an atom. The atomic number is placed to the bottom left of the symbol of the element and the mass number is placed to the top left. For example, ${}^{23}_{11}Na$ means that in one sodium atom there are 11 protons and 11 electrons. As the mass number is 23 the number of neutrons present in the atom will be $23 - 11$, that is 12.

For practice, given that the iodine atom can be represented by ${}^{127}_{53}I$, try and work out how many protons, neutrons, and electrons are present in one atom of iodine.

John Dalton believed that all atoms of a particular element were identical and this idea persisted until 1913 when Frederick Soddy challenged it. The reasoning behind his challenge is a little difficult to follow at this stage but his conclusions were so important that they cannot pass without mention. Soddy stated that it was possible for an element to have two or more different kinds of atom. Chlorine, for instance, has two kinds of atom which we can represent as ${}^{35}_{17}Cl$ and ${}^{37}_{17}Cl$. This means that both kinds of chlorine atom have 17 protons and 17 electrons the only difference between them being that one kind of atom has two neutrons more than the other. As both kinds of atom have the same atomic number they will occupy the same position in the Periodic Table. Soddy called them *isotopes*, a word made up from two Greek words, *isos* meaning same and *topos* meaning place, implying that isotopes occupy the same place in the Periodic Table. Natural chlorine as found in rock salt, etc., is made up of the two isotopes ${}^{35}_{17}Cl$ and ${}^{37}_{17}Cl$ in the ratio of approximately 3:1, so that the relative atomic mass of naturally occurring chlorine is 35.453. Isotopes of an element are very similar indeed in their chemical reactions.

It has been found that nearly all the elements have isotopes. Iodine is one element which has only one kind of naturally occurring atom, $^{127}_{53}I$. However, it is possible to make *artificial* isotopes and $^{131}_{53}I$ is one example. But this is another story to be left until later.

Radioactivity

Radioactivity is not just as old as the hills – it is probably a good deal older. But it is only comparatively recently that much attention has been paid to it; in fact man has only been aware of its existence since 1896. What then do we mean by radioactivity? The Frenchman, Henri Becquerel, working in his Paris laboratory in 1896, discovered, quite by accident, that a photographic plate, although well wrapped in black paper, was darkened as if it had been exposed to light when placed near some uranium ore. He soon realised that the exposure of the plate was due to rays emitted by the uranium. Heating and cooling did not affect the intensity of the rays. Becquerel called the emission *radioactivity* and the fact that uranium atoms gave up these rays cast doubt on the picture of the atom put forward by Dalton about 90 years previously. You will remember from Chapter 4 that Dalton believed that all matter is made up of atoms which are indivisible particles which cannot be created or destroyed.

Uranium is not the only element which is radioactive. We have only to refer to the Periodic Table to find the elements with very big atoms which provide examples of radioactive elements. Very big atoms tend to be far less stable and emit radioactivity. Elements from polonium onwards are radioactive.

It has been found that there are three kinds of radioactivity; alpha rays (α rays), beta rays (β-rays) and gamma rays (γ-rays). Let us have a word about each of these types of radiation.

Alpha rays. These consist of the nuclei of helium atoms, each alpha particle consisting of two protons and two neutrons and represented by He^{2+}. In other words they are helium atoms from which the electrons have been removed. Alpha rays are easily stopped by a few sheets of paper.

Beta rays. These consist of electrons and are more penetrating than alpha rays.

Gamma rays. These are a form of energy in the same way as heat and light are forms of energy. They can penetrate the human body and it takes several metres of concrete to stop them. These rays are used in medicine for cancer treatment.

Half-life

So far we have said that radioactive elements can give out three kinds of radiation. The particular kind of radiation depends on the element concerned, but one thing is common to all radioactive elements. The rate at which they break down and in so doing emit radiation is a definite value for each isotope. The radioactive isotope of uranium $^{238}_{92}U$, for example, emits radiation, or decays as we say, at such a rate that it takes 4500 million years for *one half* of it to change. This period of time is called the *half-life* of the $^{238}_{92}U$ isotope and is a constant value. It is not affected by temperature, pressure, the original quantity of the isotope, or any factor you care to mention.

Suppose we have 10 g of $^{238}_{92}U$. In 4500 million years there would be 5 g left. After a further 4500 million years there would be 2·5 g and so on. All radioactive isotopes, or radioisotopes as they are called, have their own half-life values. For $^{131}_{53}I$ it is 8 days; for $^{90}_{38}Sr$ it is 28 years. You may ask "Why half-life? Why not just its life?" The answer is that there is no real beginning and no real end to radioisotopes. We cannot say with any certainty when every single atom of every radioactive substance has stopped emitting radiation.

Finding out the age of materials

The element carbon is present in all living matter. Most of the carbon is made up of $^{12}_{6}C$ atoms but there are a few $^{14}_{6}C$ atoms, the proportion being one million million to one. These isotopes of carbon, which we can refer to as Carbon-12 and Carbon-14 for convenience, undergo the same chemical reactions and both are present in the carbon dioxide produced during respiration, photosynthesis, etc., but the Carbon-14 isotope differs from the other in that it is weakly radioactive, giving out beta rays, which can be detected by using a Geiger counter. In all living matter there is a fixed proportion of Carbon-14 atoms, but once the matter has died, fresh supplies are cut off and the radioactive atoms split up at a steady rate. The half-life of Carbon-14 is 5600 years; in other words, it takes 5600 years for one half of the radioactive carbon to change. From a knowledge of the Carbon-12 to Carbon-14 ratio at the time of death and measurement of the ratio of the two isotopes at a particular time, it is possible to work out how long the material has been dead. Libby of the University of Chicago is credited with the introduction of this method of *radioactive dating* as it is called.

Application of the method has led to a number of valuable and fascinating discoveries.

The age of the Dead Sea Scrolls was estimated by the method. In 1947 two Bedouin boys were playing in the caves near the Dead Sea when they came across some scrolls made of strips of leather and wrapped in linen. Their findings were tested using the Carbon-14 method and the linen, made from flax, was shown to be about 1900 years old. The scrolls have been studied by scholars of Hebrew and proved to be part of the Old Testament.

The oldest city so far discovered is Jericho. Excavations have revealed that throughout the ages buildings have been constructed on top of old ruins so that by studying the different layers archaeologists can build up a picture of successive civilisations. The bottom layer has been estimated to be about 9000 years old.

The French caves at Lascaux, adorned with wall paintings of prehistoric man indicate an age of 15000 to 16000 years. Indian sandals found in Oregon, U.S.A., are estimated at 9000 years old and there are many, many more examples of the use of Carbon-14 dating.

How old is the Earth?

Earlier in this chapter we said that radioactivity was at least as old as the hills and probably a good deal older. An estimate of the age of the Earth has been made by analysis of suitable radioactive minerals. The radioisotope $^{235}_{92}U$ in uranium-bearing rocks undergoes a rather complicated series of disintegrations, the final product being the stable lead isotope, $^{207}_{82}Pb$. Experiments to find the proportion of $^{207}_{82}Pb$ present in the uranium ore enables us to work out the age of the ore and so the age of the Earth. It is assumed, of course, that there was no lead present in the ore at the very beginning. The Earth has been estimated at 4000 million years old.

Atomic energy

A study of isotopes would be incomplete without a mention of atomic energy. It was Rutherford in 1923 who described each atom as having tremendous power awaiting release. Two German scientists, Hahn and Strassman, discovered in 1939 that when atoms of very heavy elements like uranium are struck by neutrons, splitting or fission of the big nucleus of the atom takes place. The process is called *nuclear fission* and the result is that two smaller atoms are formed from each uranium atom and a tremendous amount of energy is released. There is very little difference between the mass of the uranium atom and the total mass of the two fragments, but this change in mass accounts for energy produced.

$^{235}_{92}U$ is one isotope which will undergo fission. In addition to the two fragments, each atom gives up two or three neutrons which are capable of splitting other uranium atoms. You will see that these

Fig. 56 Chain reaction

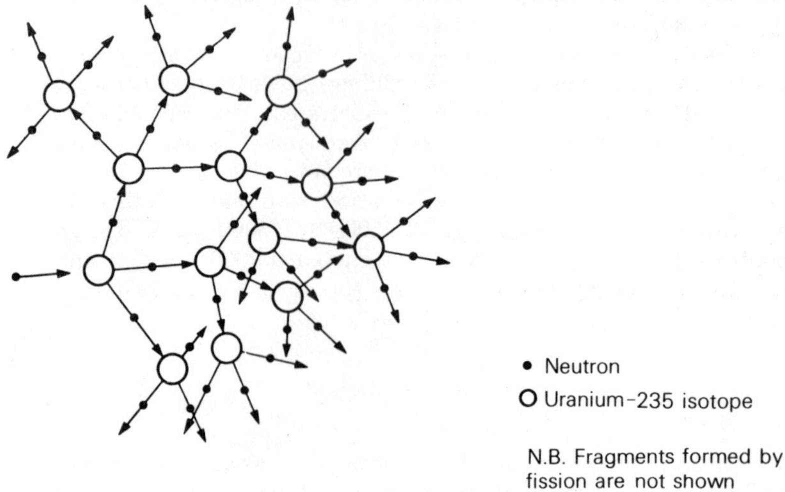

uranium atoms will each provide two or three neutrons which can cause further fission and we have what is known as a *chain reaction* setting in (Fig. 56). In the diagram, for the sake of simplicity, each uranium atom is shown to give three neutrons.

If this chain reaction is allowed to proceed without control an atomic explosion results. A tremendous amount of energy is set free in the form of heat and blast not to mention dangerous radioactivity due to radioisotopes – the fragments formed on splitting uranium atoms. This is what happens when an atomic bomb explodes.

In the nuclear reactor, however, nuclear fission is controlled using carbon which slows down neutrons, and boron rods which are inserted to absorb neutrons. The energy obtained from the atom is used to change water to steam which is then used to drive the turbines which generate electricity. For the same weight of fuel, nuclear energy releases more than a million times as much heat as chemical energy.

Radioactive dust or "fallout"

Although man has been subjected to radiation ever since he first appeared on earth, the amounts received from natural sources are so small as to be harmless. Atomic explosions produce many radioisotopes such as Iodine-131 and Strontium-90, which can be very dangerous indeed. At present there is no cause for alarm but, if the concentration of these substances in the atmosphere increases, there could be grave consequences. After an atomic explosion the radioactive dust is carried by the winds around the world and is eventually brought down by rain to the Earth's surface. Animals eat the grass which has been contaminated by the radioactive isotopes and as we eat the meat from the animals or drink their milk, the radioisotopes are transferred to our bodies. What happens then? If we take Iodine-131, for instance, it concentrates in the part of our body where normal iodine, Iodine-127, is stored. This part of the body is to the front of the neck and is called the thyroid gland. Iodine-131 emits γ-rays and these can cause serious harm to the cells of this gland. On the other hand, sometimes the thyroid gland becomes too active and the person becomes ill. This condition can be treated by administering very carefully calculated doses of radioactive iodine, which settles in the thyroid gland thus reducing the activity of the cells there.

Strontium is very similar to calcium, in fact, the two elements are in the same group of the Periodic Table. There is much calcium in bones and teeth and the Strontium-90 isotope lodges itself in these places. It is a γ-emitter and can produce very serious effects on the bones and consequently on the blood of a person, since it is in the bone marrow that the red blood corpuscles are produced.

Is there cause for worry? Not at all *provided that the proportion of radioisotopes does not increase greatly.* Actually, the amount of radiation from atomic explosions is extremely small compared with that we receive from natural sources. Things must be kept in check, nevertheless, and regular tests on the amount of radioactive matter in the atmosphere are made.

Separating uranium isotopes

Both $^{235}_{92}U$ and $^{238}_{92}U$ isotopes are present in naturally occurring uranium ores and separation of the two is essential. This has been a very difficult problem to solve since the properties of the two isotopes are so very much alike and separation of substances depends on differences in properties such as differences in solubility, melting point, boiling point, etc. One method which has been very successful depends on the process of *diffusion* (see Chapter 4). The uranium is converted to the gas uranium hexafluoride, UF_6, some molecules of which will contain the Uranium-235 isotope and the rest will contain the Uranium-238 isotope. Now these two types of UF_6 will differ slightly in density and the molecules will, therefore, diffuse at different rates, and it is this fact which allows the separation of the two types of UF_6 and therefore the two uranium isotopes. Another method is to use a special, and very expensive kind of centrifuge, called a gas centifuge, in which the heavier gases tend to move to the outside of the instrument.

Use of radioisotopes

We have already mentioned the use of the naturally occurring radioisotopes, Carbon-14 and Uranium-235, in the dating of materials. An exciting and invaluable development over the last 30 years or so has been the manufacture of radioisotopes from non-radioisotopes. Radioisotopes of almost every element you care to mention can be made by bombarding "normal", non-radioactive, elements with particles like neutrons, the conversion being carried out in nuclear reactors.

Example $^{59}_{27}Co + 1 \text{ neutron} \longrightarrow ^{60}_{27}Co$
(non-radioactive) (radioactive)

The Radiochemical Centre at Amersham, England, was established in 1946 to provide radioactive substances for both home and export uses. Today 29 radioisotopes are available in many different compounds, and these are sent to almost every country in the world. Why are radioisotopes of such importance? There are four ways in which they can be used.

1. As "tracers" because the radiation they emit can be detected and therefore the movement of the element followed.
2. As sources of radiation which can be used to kill bacteria, unwanted cells, etc.
3. As sources of energy when they are used to provide heat to generate electricity, or to produce light in luminous dials and watches.
4. In instruments which detect cracks in pieces of metal by bombarding with β or γ rays, and detecting the radiation which goes through the metal with photographic film.

Let us have a look at a few of the many applications of radioisotopes.

Industry

Cracks in castings and in welds can be found by gamma radiography. A small piece of a radioactive material (usually Cobalt-60 or Iridium-192) housed in a portable lead shield shines γ rays through the piece of metal, and cracks show up on a photographic plate. No electric power supply is needed and the equipment can be taken almost everywhere.

Radioisotopes have been used to estimate the wear of metals. In combustion engines, for example, the pistons and cylinder walls are worn during operation and it is important for engineers to know how quickly metal is worn away through friction. At one time, the only method was to take the engine to pieces and then measure the thickness of the various parts – a long and tedious job. Today it is possible to include a radioisotope in the metal parts and as the metal is worn away and passes into the lubricating oil, some of the radioactive isotope is carried with it. By measuring the radioactivity in the lubricating oil, and knowing the amount of radioactive metal present in the parts at the start, it is possible to work out the amount of metal worn away. Tyre wear can also be estimated in a similar way using a suitable radioisotope like Iodine-131 mixed with the rubber of the tyre. Leaks in underground pipes carrying oil, water, etc., can be detected by including a little radioactive solution in the liquid being pumped. Radioactivity will accumulate in the soil around the leak and the position of the leak can be detected from the road above without a lot of unnecessary digging.

Medicine

Radiocobalt (Cobalt-60) is used in cancer therapy. You may have heard of a hospital having a cobalt bomb. This is really a misnomer; it is not a bomb in the normal sense of the word. It has probably become known by this name because the radioactive substance must always be enclosed in thick layers of carbon and lead, substances which are impenetrable to radiation, when the cobalt is not in use. It is the gamma rays emitted by the Cobalt-60 which help to destroy the cancer. In another application of Cobalt-60, medical equipment such as syringes and dressings are sealed in plastic bags and passed through a conveyer system where they are subjected to gamma radiation from the radioisotope. The materials themselves do not become radioactive but all bacteria are killed or made inactive. Medical equipment sterilised in this manner is ready for use as soon as it is taken from the plastic bag, a vast improvement on the old-fashioned method of heating the equipment in boiling water for a period to sterilise. What is more, the irradiated syringes are made of cheap plastic and can be disposed of immediately after use.

One of the most recent uses of radioisotopes has been to power the heart "pace-makers" on which the lives of some people depend. Special batteries which contain the radioisotope and thermocouples are put in the pace-maker. As the radioisotope gives off radiation, some of this is converted to heat, and the thermocouples change the heat to electricity. These types of batteries last for many years.

Agriculture

Radioisotopes have been used to check how fertilisers and insecticides work. The wild silk moth has been injected with a Carbon-14 compound to study insect metabolism; research of this kind helps in the quest for new and improved insecticides.

The screw-worm fly whose grubs feed on the living flesh of cattle, has been virtually eliminated from the South-eastern states of the U.S.A. by sterilising millions of male flies with gamma rays, and releasing them. Most of the females mated with sterile males and had no offspring, and so the screw-worm fly is no longer a pest in that area.

Research

Probably every branch of science has benefited in some way from radioisotopes. Biochemists have been able to use radiocalcium to find out how long it takes for calcium in food to appear in various bones, the path of the radiocalcium, and therefore the path of ordinary calcium, being followed by some suitable device outside the body. The distribution of phosphorus in leaves has been traced using Phosphorus-32. Chemists can trace the movement of atoms during a chemical reaction by using radioactive isotopes. The future of radioisotopes is very bright indeed.

QUESTIONS TO TRY

1. What is meant by the following terms:
(a) atomic mass (d) isotope (g) nuclear fission
(b) atomic number (e) radioactivity (h) half-life?
(c) mass number (f) radioisotope

2. Complete the following word equations:
(a) Sodium + water ⟶ ... + ...
(b) ... + water ⟶ potassium hydroxide + ...
(c) Calcium + ... ⟶ ... + hydrogen

3. A piece of calcium was allowed to react with water. The resulting mixture was filtered and carbon dioxide was bubbled through until no further reaction took place. Describe and explain all the changes observed.

4. Refer to the Periodic Table and write down the formula for:
(a) rubidium chloride (b) radium sulphate (c) strontium iodide.

5. What is the valency of uranium in the compound UF_6?

6. Copy out the following table in your book and fill in the gaps.

Element	Mass number	Atomic number	Number of protons	Number of neutrons
Sodium	23	11		
Oxygen	16		8	
Calcium			20	20
$^{90}_{38}Sr$				

7. Cobalt-60 has a half-life of 5·3 years.
(a) How long will it take for exactly one half of a given sample to decay?
(b) What weight of Cobalt-60 will be left if a sample weighing 10 g is allowed to decay for 10·6 years?

Miscellaneous Questions

1. Name
  (a) a metal which easily burns in air;
  (b) a metal which forms an insoluble sulphate;
  (c) a metal which would displace silver from silver nitrate solution.

2. Name
  (a) a divalent metal;
  (b) a monovalent non-metal;
  (c) an acid which is used as a drying agent;
  (d) hydrated salt.

3. Name
  (a) a gas which burns in air;
  (b) a gas which is less dense than air;
  (c) a gas which is more dense than air.

4. What would you *see* when
  (a) carbon dioxide is led into lime water until no further action takes place;
  (b) sulphur is burned in oxygen;
  (c) concentrated sulphuric acid is left in the open for several days?

5. Describe an experiment to show that air contains about 20 per cent oxygen by volume.
Name
  (a) two other constituents of air;
  (b) two substances usually present in air over an industrial town but not in open country.
Give two pieces of evidence indicating that air is a mixture.

6. Describe what you would see and name the compound formed when a mixture of iron filings and powdered sulphur is heated. Give **one** chemical test (other than the effect of heat) by which you could distinguish between the mixture and the compound formed by heating it.

7. State whether each of the following is an element, compound or mixture. Give one reason in each case.

(a) phosphorus
(b) petroleum
(c) air
(d) water
(e) sulphur
(f) mercury oxide.

8. Describe how you could prepare and collect gas jars of:

(a) oxygen
(b) carbon dioxide
(c) hydrogen.

9. Explain how, starting from limestone, you would prepare:

(a) calcium oxide
(b) calcium hydroxide
(c) calcium hydrogencarbonate solution.

10. Among substances which occur naturally name:

(a) four gaseous elements
(b) two gaseous compounds
(c) two metals
(d) two soluble salts
(e) two naturally occurring chemical reactions.

11. Name the process used:

(a) to obtain salt from an aqueous salt solution;
(b) to obtain water from an aqueous salt solution;
(c) to separate a mixture of salt solution and sand;
(d) to separate the coloured dyes in black ink;
(e) to separate oil into its constituents;
(f) to split up a substance by an electric current;
(g) to coat an object with metal using electricity;
(h) to cause a solid to be formed immediately from a solution;
(i) to "cure" acidity by adding a base;
(j) to coat an object with zinc to prevent rusting.

12. Which colours are missing in the following?

(a) When ... mercury oxide is heated a ... gas is liberated and a ... solid remains.
(b) When ... sulphur is heated in oxygen it burns with a ... flame giving ... fumes of sulphur dioxide.
(c) When carbon dioxide is led into lime water a . . . precipitate of calcium carbonate is formed at first.
(d) When ... copper oxide is dissolved in dilute sulphuric acid a ... solution of copper sulphate is formed.

(e) When ... hydrated copper sulphate crystals are heated ... anhydrous copper sulphate remains. When water is added to this residue a ... colouration appears.

13. Name and where possible give the symbols or formulae of **three** examples of each of the following:
   (a) elements
   (b) metals
   (c) non-metals
   (d) compounds
   (e) mixtures
   (f) acids
   (g) bases
   (h) salts
   (i) types of radiation
   (j) methods of separating mixtures
   (k) particles found in atoms.

14. How many out of 50?

What is the meaning of:

1. Acid
2. Acid salt
3. Adsorption
4. Alkali
5 Alpha ray
6. Anode
7. Atom
8. Atomic number
9. Basc
10. Beta ray
11. Catalyst
12. Cathode
13. Chemical change
14. Chromatography
15. Compound
16. Covalent bond
17. Crystallisation
18. Deliquescence
19. Distillation
20. Diffusion
21. Distillate
22. Efflorescence
23. Electrolysis
24. Electrolyte
25. Element
26. Evaporation
27. Filtrate
28 Gamma ray
29 Half-life
30. Hydrate
31. Hygroscopy
32. Ionic bond
33. Ions
34. Indicator
35. Isotope
36. Mass number
37. Molecule
38. Periodic classification
39. Physical change
40. Precipitation
41. Radioactivity
42. Radioisotope
43. Reversible reaction
44. Salt
45. Solute
46. Solvent
47. Sublimation
48. Valency
49. Volatility
50. Water of crystallisation

15. Crossword on opposite page.

Across

4. Similar to a stalactite but made of water. (6)
7. .... isotopes of elements can be very useful. (3 and 5)
8. Sodium (symbol) + a noble gas used in lighting. (6)
10. When this is in FI and E it will pass through a filter paper. (5)
13. Found in palms, coconuts and underground. (4)
14. Symbol for a noble gas + symbol for tin. (4)
15. Symbols for antimony, carbon and potassium in order. (4)
16. A product of cumbustion? (3)
17. Mostly water: many eaten on a hot day. (4)
19. Manganese (symbol) enclosed in arsenic (symbol). (4)
21. The smallest parts of substances that can have separate existences. (9)
23. The source of sugar? (4)
24. A proton is said to have a .... charge of positive electricity. (4)
26. Before titrating, the part below this must first be filled. (3)
27. A mixture of radium, tungsten and iodine (symbols). (4)
29. .... is often helpful in solving any problem. (4)
32. This of decay is useful to know. (jumbled). (4)
33. Boron, sulphur, nitrogen, iodine and uranium in order. (5)
34. Chemical reactions take place when we ...... our food. (6)
35. Negatively charged and found in atoms. (8)
36. Atomic ...... somewhat upset. (6)

Down

1. Supporter of flowers and contains chlorophyll. (5)
2. Grow on trees and contain acid. (5)
3. The proton, neutron and electron can each be said to be .... of an atom. (4)
4. A type of bond. (5)
5. These are present in all solutions of electrolytes. (4)
6. The U.S. can be said to have the ...... share of (symbol for sulphur) (add the two together). (6)
9. This will neutralise an acid. (6)
11. Chemists must ... experiments to find things out. (3)
12. Goes to the positive electrode during electrolysis. (5)
13. This word could be applied to the origin of chemistry. (7)
15. Important to ... what is happening during an experiment. (3)
16. This to a problem is what we try to find (abbrev.). (3)
18. The milkiness ...... when carbon dioxide is passed into a suspension of calcium carbonate in water. (6)
20. Short for methylated spirits. (5)
21. A look at .... of the world indicates how much of the surface is covered by water. (3)
22. Sodium, potassium and chlorine are ... valent. (3)
23. This moves to the negative electrode during electrolysis. (6)
25. See 22 down. (3)
28. The commonest solvent. (5)
30 The noble gases were always called this at one time. (5)
31. Litmus ..... red in acids. (5)
32. A .... of chemists is often engaged in research. (4)
33. In a reversible reaction a .... 'ward reaction as well as a forward reaction is possible. (4)

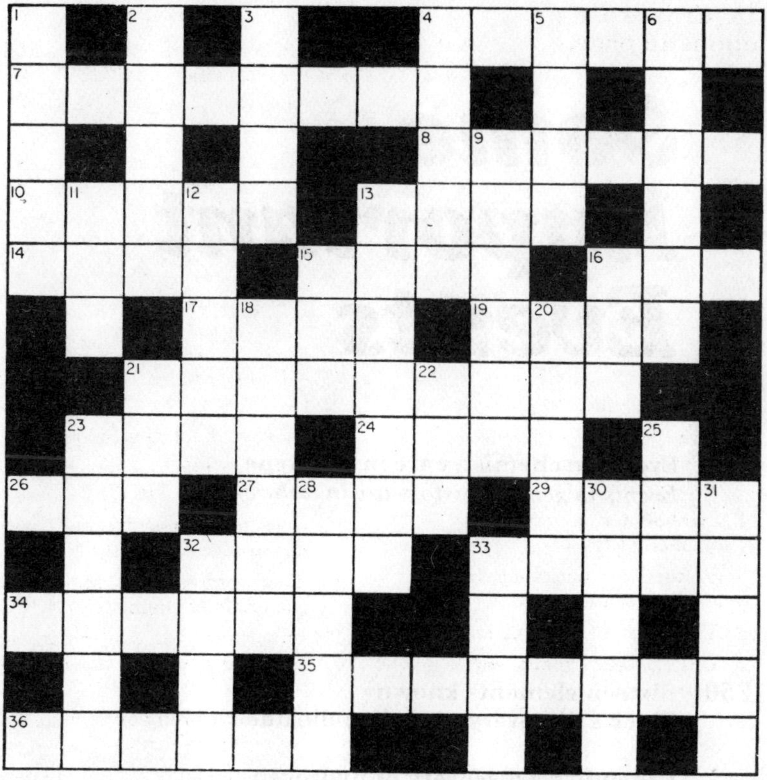

16. In each of the following groups all the members except one have one particular thing in common. State which is the "odd one out" in each group and give your reasons in full.
Example Sodium, calcium, silver and sulphur. *Answer:* All four are elements but sulphur is the only non-metal; the others are metals.

(a) Carbon, sodium, sulphur and phosphorus.
(b) Potassium nitrate, sodium hydroxide, sodium sulphate and calcium chloride.
(c) Citric acid, hydrochloric acid, sulphuric acid and nitric acid.
(d) Combustion, neutralisation, distillation, rusting.
(e) Crystallisation, filtration, chromatography and respiration.
(f) Oxygen, argon, chlorine and nitrogen.
(g) Water, petroleum, carbon dioxide, and sodium chloride.
(h) Mercury, hydrogen peroxide, lead nitrate and sodium chloride.
(i) Uranium, radium, Strontium-90, Iodine-127.
(j) Atom, molecule, proton, neutron.
(k) Lithium, calcium, sodium, potassium.
(l) Magnesium, calcium, strontium and aluminium.
(m) Alpha particle, beta particle, cation and proton.
(n) Anion, cation, neutron and electron.

Some Important Events

Events in chemistry are in bold type
Events in general history are in italic type

1750 **Sixteen elements known**
Black's discovery of carbon dioxide

Cavendish's discovery of hydrogen
Napoleon Bonaparte born

Priestley's discovery of oxygen
Lavoisier's experiments on air
Declaration of independence by the United States of America

Start of the French Revolution

1800

Davy's isolation of sodium and potassium
The Battle of Trafalgar
Napoleon's retreat from Moscow
The Battle of Waterloo
Modern atomic symbols first used

The first railway from Stockton to Darlington

Faraday's laws of electrolysis

Anaesthetics used for first time

Some Important Events

1850

First synthetic dye
The Crimean War

The American Civil War
Mendeleef's Periodic Table
Suez Canal completed

Electron discovered
Becquerel discovered radioactivity
1900 **Madame Curie discovered radium**
The first aeroplane
Noble gases discovered

The First World War
The Russian Revolution

Fleming discovered penicillin

Chadwick discovered the neutron
The Spanish Civil War
Nylon invented
The Second World War

1950 **Terylene invented**

First men in space
Structure of DNA, the basic genetic material discovered
First moon landing

1980

Index